CHROMOSOME BOTANY

by C. D. Darlington, F.R.S.

CHROMOSOMES AND PLANT BREEDING—1932
RECENT ADVANCES IN CYTOLOGY—1932, 1937
THE EVOLUTION OF GENETIC SYSTEMS—1939, 1958
THE CONFLICT OF SCIENCE AND SOCIETY—1948
THE FACTS OF LIFE—1953
THE PLACE OF BOTANY IN THE LIFE OF A UNIVERSITY—1954
DARWIN'S PLACE IN HISTORY—1959
GENETICS AND MAN—1964
CYTOLOGY—1965
THE EVOLUTION OF MAN AND SOCIETY—1969

Jointly

THE HANDLING OF CHROMOSOMES—1942, 4TH EDN 1962
(with L. F. La Cour)
CHROMOSOME ATLAS OF CULTIVATED PLANTS—1945
(with E. K. Janaki Ammal)
THE ELEMENTS OF GENETICS—1949
GENES, PLANTS AND PEOPLE—1950
(with K. Mather)
CHROMOSOME ATLAS OF FLOWERING PLANTS—1956
(with A. P. Wylie)
TEACHING GENETICS—1963
(with A. C. Bradshaw, ed.)

ORIGINS OF CULTIVATED PLANTS

Ipomoea°

Zea°
Maranta
Agave
Psidium
Helianthus

Ipomoea°
Phaseolus°*
Capsicum*
Cucurbita*
Gossypium 4x*

Solanum°*
Lycopersicum
Canna
Nicotiana*
Cinchona

Manihot°
Arachis°
Carica
Ananas
Anacardium
Theobroma
Hevea

Vicia°
Ceratonia
Asparagus
Lactuca
Humulus

Nuclear
Zone crops

Basic
Triticum°*
Hordeum°*
Pisum°
Lens°
Cicer°
Linum°

Secondary
Olea
Vitis
Ficus
Allium*
Brassica*
Beta

Ribes
Rubus
Avena°
Secale°

Prunus
Pyrus

Cucumis*

Fagopyrum°
Setaria°
Cannabis

Phoenix°
Punica
Morus

Panicum°
Cucumis*
Corchorus*

Glycine°
Morus
Thea

Saccharum°
Oryza°
Piper

Eleusine°
Guizotia
Coffea

Cinnamomum

Colocasia°
Dioscorea°
Musa°
Cocos°
Citrus

Elaeis°

Sorghum°
Sesamum°
Ricinus

Gossypium 2x°

N. Z.

POLYNESIA

○ Basic crops
● Several species

C. D. Darlington

CHROMOSOME BOTANY
and the
Origins of Cultivated Plants

THIRD (REVISED) EDITION 1973

London
GEORGE ALLEN & UNWIN LTD
RUSKIN HOUSE MUSEUM STREET

FIRST PUBLISHED IN 1956
SECOND EDITION 1963
THIRD (REVISED) EDITION 1973

ISBN 0 04 581011 7

PRINTED IN GREAT BRITAIN
in 11 *point Plantin type*
AT THE UNIVERSITY PRESS
ABERDEEN

To
OTTO RENNER 1883–1960
Professor of Botany in the Universities of
Jena and Munich
who discovered the nature of hybrid species

The standing objection to botany has always been that it is a pursuit that amuses the fancy and exercises the memory without improving the mind or advancing any real knowledge; and where the science is carried no farther than a mere systematic classification, the charge is but too true. But the botanist that is desirous of wiping off this aspersion should be by no means content with a list of names; he should study plants philosophically, should investigate the laws of vegetation, should examine the powers and virtues of efficacious herbs, should promote their cultivation, and graft the gardener, the planter, and the husbandman, on the phytologist. Not that system is by any means to be thrown aside; without system the field of Nature would be a pathless wilderness; but system should be subservient to, not the main object of, pursuit.

GILBERT WHITE

Letter June 2, 1778

Another result of the depreciated state of systematic botany is that intelligent students, being repelled by the puerilities which they everywhere encounter, and which impede their progress, turn their attention to *physiology* before they have acquired even the rudiments of classification or an elementary practical acquaintance with the characters of the natural orders of plants.

JOSEPH HOOKER

Flora Indica 1855

PREFACE

OUR studies of plants are more and more concerned with their existence in societies. In societies we see them in relation to one another, to the earth on which they live and to the changes that they are continually undergoing, changes which together constitute evolution. Yet when we consider plants in societies we find that the two conditions in which they live, as wild plants and as cultivated plants, are so contrasted as to split the whole science, so contrasted that it is hard to get one man or even one school to study both. A division has grown up between schools which study the pure botany of wild plants and those that study an applied or economic or agricultural botany of cultivated plants. This cleavage is scarcely lessened by the formation of yet a third school which studies the cultivation of wild plants. All this is natural and some of it is necessary. But it is also natural and necessary that the information acquired by each of these schools should be allowed to fertilize the others.

The same cleavage is found between experimental breeding and chromosome study which together make the foundation of genetics. The two methods are so highly contrasted that few investigators can be expected to practise both. Yet they are mutually indispensable. This is not only because they arise from and give rise to the same theory, which fortunately no one doubts: it is also because they use different kinds of material. Experimental breeding is concentrated on a few hundred species, most cultivated species. Chromosome studies are diffused over many thousands of species, mostly wild species. Each therefore is a check on the other. Both are a check on the descriptive methods which have, and will always have, a wider range than any of the analytical and experimental methods of genetics. Above all they make it possible to bring our knowledge of wild and cultivated plants, our knowledge of systematics and ecology into one focus. To achieve this end is the purpose of the present book.

The difficulty of the book arises from the connection of chromosome studies with genetic principles, a connection which is reciprocal and therefore not a little intricate. I have, however, avoided this intricacy by treating the matter to a large extent descriptively, and by pointing to the problems as much as by providing solutions.

The first chapter is an introduction to the use of chromosomes for botanical studies. The second deals with their relations with classification; the third and fourth with their relations with plant geography and ecology and evolution. In the last three chapters the study of cultivated plants is shown to confirm and sometimes to clarify the same succession of ideas as arise from the study of wild plants. Part of the fifth chapter is based on the *Introduction to the Chromosome Atlas of Cultivated Plants*. The new and enlarged version of this work, the *Chromosome Atlas of Flowering Plants*, will be found useful by those who wish to pursue further the problems of plant systematics, plant geography and evolution.

Now the chromosomes underlie not only the whole of botany but the whole of biology. There is no distinction in general physiology and genetics between what they do in plants and in animals. Indeed it is vital in these respects to consider plants and animals together. Why therefore should one take the chromosomes separately in relation to plants, and even to a special group of plants?

The answer is that the flowering plants are better understood than any other large group of plants or animals in regard to their ecology, their geography, their systematics, their breeding behaviour and general genetics, and, above all, the evolution of their chromosomes. Man's greatest biological experiment has been the invention of agriculture, a process of understanding and controlling and improving certain flowering plants. In the flowering plants therefore we may say that these properties have been studied with a concentration and an effect altogether greater than in any other group of organisms.

The fruits of this study can, I believe, be enjoyed by those concerned with plants and with animals in nature, and even by those who are concerned with them in a radically different way. For this reason I am greatly indebted to my colleague Dr E. B. Ford for pointing out, amongst other things, the bearings of this work on the wide range of evolutionary problems with which he is familiar. His comments are included in an appendix.

Here then is the field of Chromosome Botany. It is one of the several aspects of botany which is fundamental for the study of life as a whole.

Botany School,   C. D. DARLINGTON
University of Oxford

PREFACE TO THE THIRD EDITION

IT is now thirty years since I prepared the first sketch of this book. During that time the growth of genetic and chromosome studies has shown the evolution of the Flowering Plants, with which the book is mainly concerned, in an entirely new perspective, new in a sense which has become fundamental for the study of evolution as a whole. To explain this is one of the aims of this new edition.

During that time also the parallel growth of many sciences has brought the question of the origins of cultivated plants from the edge of enquiry into the very centre of human affairs. Above all we now know (although some archeologists may forget it) that civilization has always been the work of men who grew grain crops and lived on them. Since we also know (partly by their chromosomes) what wild grains they first grew, we also know where to find the origins and how to trace the movements of civilization. To explain these things is the other aim of this new edition.

C. D. DARLINGTON

CONTENTS

ILLUSTRATIONS

LIST OF TEXT FIGURES

TABLES

I. CHROMOSOMES

I. CELLS AND NUCLEI

PLANTS and animals can be seen under the microscope to be divided into compartments or cells. But each individual plant or animal, each organism, arises from the growth of an egg which had no such compartments. It was a single cell containing a single nucleus. The change or development depends on the repeated division of the original cell into pairs of daughter cells. At each successive division the nucleus divides along with the cell, a partition growing across the cell between the two daughter nuclei so that every daughter cell also comes to contain a nucleus.

At the beginning of this process of cell-division, which is known as *mitosis*, the mother nucleus resolves itself into a mass of coiled or twisted threads. Each thread is already split lengthwise into halves. These double threads are the *chromosomes* and their sister halves are the *chromatids*. Each chromatid consists of a coiled fibre which stains with basic dyes. At the middle stage, or *metaphase*, of mitosis the chromosomes arrange themselves, to the number of two or two thousand as the case may be, in a circular zone which is flat, if the chromosomes are not too long, and is therefore known as a plate. From this plate a tapering cylinder of parallel fibres is organized pointing towards opposite sides of the cell. This is the spindle. Suddenly and at once the sister chromatids fall apart. They become, as we call them, daughter chromosomes which pass towards opposite ends or poles of the spindle to form two separate groups and from these two groups the daughter nuclei are constituted.

The spindle is the product of activity of a special particle at the limit of visibility on each chromosome. This particle, the *centromere*, which, unlike the rest of the chromosome, is still undivided at metaphase, organizes or secretes or polymerizes the spindle fibres. It is only when the spindle has grown up from them that the centromeres divide or explode, their halves moving apart, passing as though along the channels which their fibre development has prepared. A centromere is thus vital for the propagation of each chromosome and the centromeres taken together are vital for the propagation or division of the cell (Fig. 1).

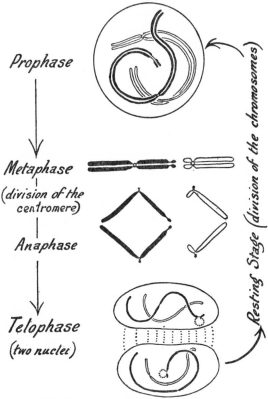

FIG. 1.—Diagram showing what happens to two sister chromosomes in the mitotic cycle. Note the nucleolar and centric constrictions (after Darlington 1932).

By mitosis, we see that two nuclei are formed containing exactly the same number and form of chromosomes as the mother nucleus. Not only this, as mitosis proceeds the whole plant (or animal) comes to contain millions of cells all with their own nuclei and each nucleus with the same number of chromosomes. In the normal course of development the whole individual has a uniform outfit or complement of chromosomes in all its cells. This constancy of the chromosomes implies (as was recognized in the nineteenth century) that they maintain their existence as coherent molecules in the resting nuclei from one mitosis to the next.

The constancy of chromosomes and nuclei is responsible for, or as we may say *determines*, the constancy with which a vegetatively propagating plant maintains its character through whole continents of space and whole epochs of time. If some accident happens to a cell a change takes place in this determined or genetic character. The failure of separation of two nuclei may lead to the formation of a new line of cells with a doubled chromosome number. These will be giant cells and, if they replace the normal type, they will form giant plants: or, if both halves of a chromosome pass to one pole at mitosis, daughter nuclei are formed with one too many and one too few. The first will be different in genetic character from the parent nucleus; it will be unbalanced. The second will be more different; it usually dies indicating that it is in a sense defective and that the whole of the chromosomes are necessary for the life of the cell.

Such accidents are not uncommon. The great constancy of nuclei and chromosomes in plants is therefore evidently due to the fact that anything different from the ordinary usually fails to survive. The mechanism of mitosis is good but natural selection, favouring old kinds of cells at the expense of new kinds of cells, makes it appear even better than it is.

2. WHAT ARE THE CHROMOSOMES?

The nucleus, it is often said, *contains* the chromosomes. But it is more correct to say that the nucleus *is* the chromosomes. The nucleus is the form in which the chromosomes exist between mitoses.

In the resting nucleus each chromosome exists as a thread

3

which has to coil up, reducing its length to one-fifth or one-tenth, to form the metaphase chromosome, and at the same time losing some of the water attached to it. The thread itself is believed to be a bundle of two, four, eight or more unitary threads. Each of these is composed of a double helix of desoxyribose nucleic acid (DNA) with which polypeptide chains are variably combined. The coiling of the helix varies cyclically in the course of mitosis as the chromosomes shorten by coiling or lengthen by uncoiling. It also varies along the length of the chromosome according to its metabolic activity to give a grouped or beaded structure.

Now the chromosome thread is liable at certain times, or even by accident at uncertain times, to break; it always does so between groups and not within groups. Each group, therefore, behaves as a unit which can be separated from other groups in the long-term properties of life, that is, in heredity and evolution. Such a separable unit or particle is known as a *gene* and the chromosome is therefore, in a genetic sense, a string of genes.

Genes often occur in groups having similar actions. The centromere is one such group governing the formation of the spindle and the movement of each chromosome on it. There is also often another structure which at the end of each mitosis secretes or organizes a *nucleolus*, a spherical body storing nucleic acid and proteins inside the resting nucleus. Nucleoli sometimes arise at the ends of chromosomes, perhaps without specific organizers. In each chromosome set, however, as a rule, there is one organizer. The position of the centromere and likewise of this organizer is still marked by an uncoiled segment or *constriction* in a metaphase chromosome.

A third kind of gene-grouping characteristic of many species of plants and animals consists of very small genes or *polygenes* having a less complicated and less specific action. These are sometimes recognizable: in the resting nucleus they remain coiled and concealed by a thick coat of sticky nucleic acid and at metaphase they may suffer from the reverse condition of nucleic acid starvation and under-coiling. They are known as *heterochromatin* or H-segments (Fig. 2).

4

The ordinary working genes of which the bulk of the chromosomes, or the *euchromatin*, is composed undergo the characteristic cycle of changes in mitosis by which chromosomes are recognized. In this cycle the chromosome loses water and its structure becomes visible as it coils and assumes a rod shape at metaphase of mitosis. And secondly, it uncoils, gains water, and ceases to be separately visible as it reverts to form the resting nucleus.

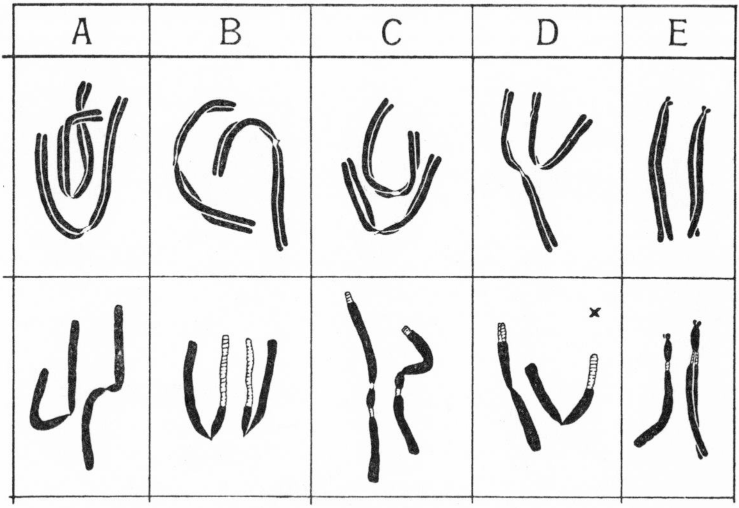

FIG. 2. The five pairs of chromosomes at mitosis in the root-tip of *Trillium stylosum:* above, normal at 20° C.; below, with nucleic acid starvation of the specific heterochromatic segments at 0° C. The cross marks an unequal or heterozygous pair of segments. × 1000 (after Darlington and La Cour 1941).

The ordinary working genes give the chromosome its general character. Its individual character which distinguishes one chromosome from another and also distinguishes each of the 3 or 4, or 10 or 20 chromosomes of a particular plant's complement from one another, and distinguishes one plant's complement from another under the microscope, this character is given by the position of the three special gene groups: these are the centromere, and in those chromosomes that have such things, the nucleolar organizer and the heterochromatin.

This character of the chromosome, as we saw, it maintains from one mitosis to another. During its invisible phase, in the resting nucleus, the chromosome therefore propagates itself. It does so by chemical processes which we can infer in some detail. The bundle of nucleic acid and protein chains of which each chromosome consists must attract identical materials to itself, chain by chain and gene by gene. This is the ultimate process of reproduction, self-propagation, and heredity underlying all others. As a rule the new mitosis does not begin until

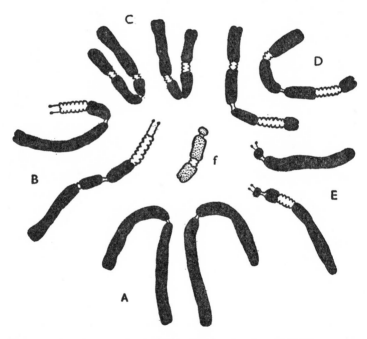

FIG. 3. Metaphase of mitosis in the root-tip of a plant of *Trillium ovatum* kept at 0°C. for four days before fixation. The segments of heterochromatin, twelve in number, and present in all but the A chromosomes and one of the E's, are starved of DNA. They correspond in the C and D pairs but B and E are heterozygous in this respect. B and E also have nucleolar satellites. There is a supernumerary chromosome, labelled f, with a heterochromatic segment.
× 1100 (from Darlington and Shaw 1959).

double the materials of the old chromosomes have been collected. When it begins, the double bundles can split into two halves, two chromatids each the size of the old bundles. The chromosomes thus remain stable in size as well as in their whole genetic content and character.

It is worth noting that the crucial property of heterochromatin probably consists in its propagating itself in the resting nucleus *later* than euchromatin. That is perhaps why at low temperatures in some plants and animals it may be brought into mitosis at half-thickness or as we say starved of nucleic acid. With fluorescent dyes heterochromatin may also be shown to have a different reaction from euchromatin. For reasons we do not know, its affinity may be greater or less, the reaction being characteristic of the species if not of the race or variety.

3. SEXUAL REPRODUCTION

In vegetative life, as we saw, changes in the chromosomes are not uncommon but are rarely perpetuated. Merely adding or subtracting a piece of chromosome is not likely to give a nucleus and its cell a better chance in life. With sexual reproduction, however, new arrangements may arise which have a better chance in life. In all plants reproducing sexually two germ cells or gametes fuse. These come from similar parents, so similar that, as a rule, they carry or deliver the same number of chromosomes into the mitosis of the fertilized egg or zygote.

It was indeed the quality of the nuclei in the mating germ cells, as contrasted with the inequality of the materials outside them, the cytoplasm, which first indicated that the chromosomes must be responsible for heredity, responsible, that is, for the long-range determination of the character of plants and animals. For in heredity, in almost all organisms, the contributions of the two parents are equal.

Now, if the number of chromosomes in the gamete is n, the number in the zygote must normally be $2n$, and the zygote can produce gametes in its turn only by an intervening process of reduction at some cell division in the course of development.

In the flowering plants this reducing mitosis, or *meiosis*,

7

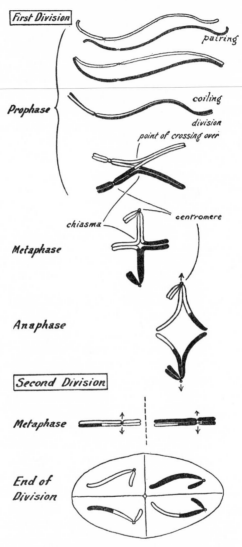

FIG. 4. Diagram showing what happens to two mating chromosomes in the course of meiosis (after Darlington 1932).

takes place in the course of formation of the pollen and the eggs. The chromosomes from the mother and father which have taken no special note of one another throughout development now demonstrate their character and relationship: they come together in pairs. Each chromosome then appears as a single string of particles corresponding more or less closely to the string of genes revealed by breeding experiments. These strings pair side by side, particle by particle. They then divide into halves as they would in a normal mitosis. As they divide the partners fall apart. At the same time something else happens to them, an event invisible in itself but with momentous consequences for the cell and for all its posterity. This event is known, by its consequences seen in breeding experiments, as *crossing-over*. By its consequences seen in the cell it is known as a *chiasma*, an exchange of partners between what are now the paired halves of the chromosomes. Any pair of chromosomes may have one or more crossings-over, one or more chiasmata, distributed along their length. The condition of the chromosomes at this time may be imitated by a double thread of wool. Twist it to give a right-handed coil. Relax the twisted thread, allow it to coil round itself. There are now four threads in association. The double threads represent chromosomes, the single threads chromatids. The chromosomes are coiled round one another in a right-handed direction, the chromatids in a left-handed direction. The two coilings are thus opposite and in equilibrium.

With this model we can see what happens in crossing-over. The split halves of the chromosomes break at one or more places at the same time that they arise by splitting. Being paired and being, moreover, coiled round one another under torsion, the chromosome-halves break in pairs and uncoil thereby releasing their own torsion. Thus, when the breaks rejoin, we find that the chromosomes have exchanged parts and crossed over and partly uncoiled. They have also at the same time fallen apart so that the position of exchange or crossing-over is marked by the visible exchange or chiasma.

Since the chromosomes are now repelling one another it is only the existence of the exchanges, the chiasmata, which holds them together in pairs. These pairs are the bivalents which we see at metaphase of the first division of meiosis. They are present in the halved or haploid number. Thus $2n$ chromosomes at mitosis give n bivalents at meiosis.

At this first metaphase of meiosis the two centromeres of each bivalent are co-oriented in the axis of the spindle as though repelling one another towards opposite poles. The separation of the chromosomes to the poles continues this movement of repulsion and arises simply from the lapse of association of the pairs of half chromosomes.

Like the chromosomes as a whole, the centromeres divide later than in an ordinary mitosis. They are still single when they pull the already divided partners apart at the first division of the meiotic cell. They divide only at the second division (Fig. 4).

There are thus two divisions of the meiotic cell and of its nucleus but only one division of its chromosomes. The double, or diploid, or $2n$ number is reduced to the single or haploid, or n, number appropriate to germ cells whose fusion will restore the diploid number characteristic of the vegetative cells.

This alternation of haploid and diploid phases in development, which underlies the alternation of generations in the lower plants, is telescoped in the higher plants; and in the Angiosperms the haploid pollen grain nucleus undergoes only two mitoses before the nuclei of the male gametes or sperm cells are formed; and the embryo-sac undergoes only two or three, or four mitoses before the female gamete or egg cell is formed. The union of the egg and sperm nuclei is known as fertilization. From the product of fertilization the new diploid embryo, the new individual, develops. There is thus an alternation of a short haploid generation, the gametophyte, and a long diploid generation, the sporophyte.

The genetic importance of meiosis is much more profound than such arithmetical or even developmental facts indicate. The paternal and maternal chromosomes which have paired have been distributed at random. If there were differences between the partners of each pair all these differences can thus be recombined in different ways. And with n pairs 2^n different combinations can be formed. But this is not all.

The crossing-over between the partners may occur once or twice or even five or ten times in each pair of chromosomes. It may occur between any two successive genes and, in every cell, the numbers and positions of crossing-over for corresponding pairs are peculiar to that cell. No haploid cell produced by meiosis is therefore likely to have the same combination of parts of maternal and paternal chromosomes as any other haploid cell: every spore or germ cell is unique. And every gene has its own history.

4. THE CONTROL OF MEIOSIS

Meiosis with the same general character occurs in all sexually reproducing organisms. That is to say crossing over and chiasma formation, chromosome pairing, segregation and reduction, follow a switch in the time relations of changes in the cell and in the nucleus (Table 1). These relations however vary in the amount of time available for pairing so that it may be complete or it may be interrupted. Pairing also may begin at the ends of the chromosomes or near the centromeres. If then it is interrupted, crossing over and chiasmata are localized near the ends or near the centromeres (Fig. 5).

These variations have certain special consequences that we shall see later. But they have one general effect that governs all heredity in all sexually differentiated plants and animals. It is that owing to the difference of size of the male and the female meiotic cells the times required for pairing are different and this makes it possible for such organisms to run their heredity along two tracks: crossing-over is distributed differently on the two sexual lines. In general there is more crossing-over on the female side in hermaphrodite plants and there is more holding together on the male side.

The extreme example of this contrast is that in which a new form of meiosis has been developed without any crossing over or chiasmata at all. The chromosomes attract one another in pairs and after splitting their chromosomes continue to attract one another in fours until metaphase when the paired centromeres separate and pull them apart. This type of meiosis with

TABLE 1

THE TIMING DIFFERENCES BETWEEN MITOSIS AND MEIOSIS

Phases	MITOSIS		MEIOSIS	Phases
R.S.	Split ↓ Double	CHROMO-SOMES	single ↓ pairing ↓ paired	PRO:— Zygo- Pachy-
PRO.		split : c.o. : Xta		
	Chromosomes separate	CHROMA-TIDS held in pairs	Chrs. held in pairs by Xta	Diplotene
MET.	Self-oriented	CENTRO-MERES	Co-oriented (repelling)	M I
ANA.	Fall apart	CHROMA-TIDS	Fall apart	A I
				INTERPHASE
	Split	CENTRO-MERES	Split	M II–A II
CELL and Chrs.	Divide once and in phase		Cell begins earlier and divides twice	Hence reduction and recombination

Chr: Chromosome. R.S.: Resting stage. M and A: Metaphase and ana-phase. I and II: First and second divisions.

the accompanying absence of crossing-over was first discovered in males of the fly *Drosophila*. It was later shown to be wide-spread in short-lived animals. It has now been found (nearly sixty years later) in the flowering plants, in two species of *Fritillaria*: there the pollen mother cells have meiosis without chiasmata; but the embryo sac mother cells have normal meiosis with chiasmata ensuring that recombination of genes is maintained in the species.

It may be that such *achiasmate* meiosis with its extreme two-track contrast in heredity is widespread in plants whose chromosomes have been too small for close study. But it is likely that it will be found only in plants with a short sexual cycle among which too much genetic recombination can be a disadvantage.

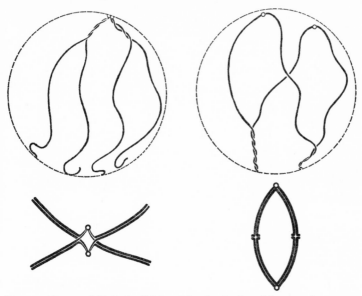

FIG. 5. Types of Meiosis: With pairing beginning, *left*, near the centromere, *right*, near the ends to give, *below*, metaphase bivalents with chiasmata, *left*, proximally localized and, *right*, distally localized.

Meiosis, combined with fertilization, is thus the means by which the hereditary materials are recombined. This is the point of sexual reproduction. In experiment it provides the basis of the chance recombination among hereditary differences which Mendel discovered after crossing or hybridization; and in nature it provides the means by which an inter-breeding group or species can expose these recombinations to the natural selection which Darwin assumed to be the agent of evolutionary change. Going still further, we see that this habit which corresponding chromosomes have of meeting and exchanging parts is what keeps a species together as a breeding unit. As soon as a barrier arises to prevent the exchange of genes between corresponding chromosomes, these chromosomes have become

13

separated in the line of evolutionary descent. At that moment we have the beginning of discontinuity and what happens in these circumstances is something we now have to consider.

5. CHROMOSOME CHANGE

It was formerly believed that the chromosome number was uniform and constant not merely for the whole of an individual but also for the whole of a species. As observations became more accurate and more extensive, however, exceptions multiplied and it became evident that the chromosomes would offer us no such simple definition of a species, no such direct corroboration or correction of the judgments of systematists, and no such simple guide to the classification of plants and animals. They were not just another 'character' to be added to the description. Indeed how could they be when they themselves determined the whole character of the species?

What is appropriate and even characteristic in the way of the fusion of germ cells does not always happen in real life. Hybridization may occur. Dissimilar germ cells may unite; and they may be derived from the same parent (on self-fertilization) or from different parents (on cross-fertilization). In these circumstances a flood of variations may ensue and every one of the four cells produced from each of the innumerable mother cells may be different from every other, not only historically but often visibly, in respect of their chromosomes.

How do these differences ultimately arise? The kinds of changes which chromosomes undergo are revealed in three ways. (i) We may see the evidence of changes which are occurring, or have just occurred, at mitosis, changes which may be spontaneous or may have been induced by special physical or chemical treatment; or (ii) we may see chromosomes which show a common origin by pairing at meiosis but nevertheless differ in the arrangement of their genes; they therefore make an asymmetrical pair; or (iii) we may compare the complements of related plants or species or even families, and thus infer changes that must have occurred in one or other of them since their lines of descent diverged.

These three methods of inference give us a consistent and, taken together, what seems to be a complete picture of how the chromosomes change and evolve.

The first kind of change a chromosome complement may undergo is in the number and proportions of its members. Beginning with the common type of diploid or $2x$ complement we may have tetraploids and octoploids ($4x$ and $8x$), usually giant plants, by doubling (Fig. 6). We may also have triploids and hexaploids by hybridization of the new and old forms. These polyploid plants may lose chromosomes ($3x - 1, 4x - 1$, $4x - 2$, and so on). Such a loss might be fatal to a diploid since all the chromosomes in the right proportions are necessary, as a rule, for the life of every cell. But in a polyploid the unbalance may be too little to cause trouble. Or, in rare cases, it may be useful: it may be adaptively suitable for special new conditions.

FIG. 6. Metaphase plates from a root-tip of *Chenopodium album* with the normal eighteen and with thirty-six and seventy-two chomosomes, resulting from doubling and redoubling at mitosis. × 1900 (from Maude 1940).

The second kind of change a chromosome complement may undergo depends on the breakage of individual chromosomes: the spontaneous breakage at uncertain times to which we have referred. As a result of some chemical disturbance or some error of reproduction, the thread breaks into two parts. One part has the centromere, the other has no centromere. The *centric* fragment remains with the nucleus if the nucleus survives the change. The other part, the *acentric* fragment, is lost at mitosis and disappears in the cytoplasm.

Sometimes the chromosomes may break at two points in the same nucleus and at nearly the same time. Then something quite new happens. The four broken ends may join up again: either as they were or in a new way. It is easy to work out in how many new ways this may happen. It rests on whether the two breaks are in one chromosome or in two, whether the result is:

(i) an *inversion* of a segment in the one (*abcd* becoming *acbd*), or

(ii) an *interchange* of segments between the two (*AB* and *CD* becoming *BC* and *DA*).

The structural changes arising in this way give rise, in all the lines descended from the changed cell, to *structural hybridity*, either inversion or interchange hybridity. In these lines the pairing of chromosomes and their crossing-over at meiosis is upset. The germ cells they produce will sometimes die and the hybrid cells or tissues will be partly sterile; or they will contain entirely new combinations of chromosome parts and even new chromosome numbers that will live and multiply.

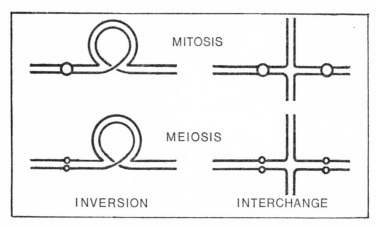

FIG. 7. The consequences of the two main types of structural change. *Above*, at metaphase following exchange between chromatids in the resting stage. *Below*, with the changed and unchanged chromosomes paired in the resultant hybrid at the prophase of meiosis.

16

To the upsets in meiosis which result where there is an odd number of chromosomes, or an odd arrangement of genes, as in an inversion or interchange hybrid, is largely due, on the one hand, the sterility of some hybrids and, on the other hand, the production of further, secondary, changes from those hybrids that are not quite sterile.

The simplest of them is the breakage of chromosomes which follows from crossing-over at meiosis between relatively inverted segments in hybrids. We find that inversion-hybrids often arise from cross-pollination not only between but also within species. The centric fragments of the broken chromosomes may then arrive in germ cells which already have normal chromosomes of the same kind. Such fragments may be very small, too small to cause a violent and undesirable change in genetic character. In this way the haploid complement may be increased. From $n = 5$ we may get $n = 6$, $n = 7$ and so on.

Fig. 8. The standard complement of 15 mitotic chromosomes in *Spiraea filipendula*: there are two telocentric chromosomes derived from splitting of one at the centromere and the species is always hybrid in respect of this change. × 3600 (after Maude 1940).

Even simpler is a kind of change which is now known to be very frequent. Unpaired chromosomes at the first division of meiosis, having no partners to separate from, and having centromeres which will not be ready to divide until the second division, may be stimulated by the spindle to *misdivide*. That is to say, their centromeres split crosswise instead of lengthwise and thus break the chromosome into two halves which go to opposite poles. Each arm of the old chromosome thus becomes a new independent chromosome with a terminal centromere, a *telocentric* chromosome. In this way also *n* may be increased. In *Campanula persicifolia* this has been observed experimentally and it may be inferred elsewhere, e.g. in *Alisma plantago-aquatica* and in *Spiraea filipendula* (Figs. 8 and 37).

In a few plants (and animals) various somatic chromosome numbers are found in different individuals or races of the same species. For example, in *Lycoris aurea* (of the Amaryllidaceae) a V-shaped chromosome breaking at the centromere, no doubt by misdivision, forms two rod-shaped chromosomes and thus increases the basic number from six to seven. Plants are found with 12 and 14 chromosomes and also the breakage-heterozygotes with 13 chromosomes.

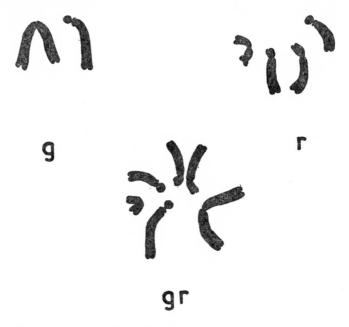

FIG. 9. Chromosome sets of *Haplopappus gracilis* (*n* = 2), *H. ravenii* (*n* = 4) and their F₁ hybrid (2*n* = 6) × 1700 (after Jackson 1962).

Such changes in number, however, will mean no change in balance unless they lead to secondary effects. This, however, they often do. For the telocentric chromosome is not always stable. Its centromere is weak; it fails to divide and the undivided chromosome passing to one daughter nucleus forms a new chromosome with identical arms, an *iso-chromosome*, which

18

again means something new in shape and in balance as well as in number.

In *Nicandra physaloides*, in addition to nine pairs of ordinary chromosomes, a pair of iso-chromosomes are found and although one of them is liable to be lost in a proportion of cells at meiosis, the other remains an indispensable and permanent member of the chromosome set. There are plants with 20 and also with 19 chromosomes. Seeds with 20 chromosomes germinate first while those with 19 germinate in nature many years later thus giving the species the capacity to survive long unfavourable periods.

6. POLYPLOIDS

In the ordinary alternation of generations between haploid gametophyte and diploid sporophyte plants we refer to the chromosome numbers as n and $2n$. And in most plants the set of n chromosomes represents an indivisible or *basic* number. In polyploids, however, there are three, four, five or more of these basic sets. $2n$ and n then become merely conventional symbols for the zygotic and gametic numbers. What interests

TABLE 2

THE EFFECTS OF ERRORS IN THE SEXUAL CYCLE

		MEIOSIS	
		+	−
FERTILIZATION	+	Sexual Reproduction $2x$ (originally)	Polyploidy $3x$, $4x$, etc.
	−	Haploid Parthenogenesis[1] x (sterile)	Apomixis: Asexual or Subsexual Reproduction[2] (usually polyploid)

Note: [1] now known in two long-established perennial clones: *Pelargonium*, Kleiner Liebling *c.* 1860 ($n = 9$, Daker 1967); *Thuja gigantea gracilis*, 1896 ($n = 11$, Pohlheim, 1968).
[2] for classification, see Darlington 1965.

19

us rather is the relation with the basic set which we may label x. In a polyploid series the 'diploid' or $2n$ number is then $3x$, $4x$, $5x$ and so on to $16x$ or $22x$.

In all species of flowering plants which normally undergo sexual reproduction either meiosis or fertilization may fail. If fertilization fails the progeny are haploid and are as a rule few, feeble, and infertile. But if meiosis fails, diploid gametes are formed and $3x$ or triploid progeny result. On the other hand, a mere failure of the nuclei to separate at mitosis in any plant gives rise to a $4x$ or tetraploid cell from which a tetraploid plant may arise by bud mutation.

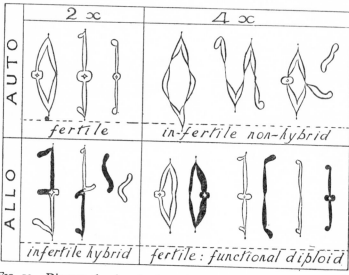

FIG. 10. Diagram showing the dependence of fertility on what happens at meiosis. Above is the situation in *Datura* or *Lycopersicum*, below that in *Primula kewensis*. Note: sterility arises *either* from irregular segregation of like chromosomes *or* from regular segregation of unlike chromosomes. In the autotetraploid all but one of the chiasmata are shown as terminal following movement of interstitial chiasmata to the ends.

Triploids with three sets of chromosomes can arise from diploid parents through a failure of reduction, or meiosis, in the formation of the parental pollen or eggs. In such plants the 3 chromosomes of each kind usually associate and separate to opposite nuclei, two-and-one whole chromosomes at the first division, or two-and-one halves at the second. Thus the separation is unequal and germ cells are formed with all chromosome numbers between x and $2x$. Most, being unbalanced, die and the triploid is therefore partly sterile: the larger the basic number the more sterile it is.

The position of tetraploids, or indeed any other polyploids with even numbers of chromosome sets, is more hopeful. Tetraploids, however, are of two kinds. Those which arise by doubling in a true-breeding or non-hybrid diploid are said to be *auto*-tetraploid. All 4 chromosomes of each kind associate at meiosis. These fours usually divide evenly into 2 and 2; but they may miss their way and divide into 3 and 1. Auto-tetraploids are thus less fertile than their diploid parents (Fig. 10).

bungei

oreades

crocea

FIG. 11.   The gametic chromosome sets in three species of *Crepis*; above two diploid species, below the tetraploid which is supposed to have arisen from their crossing. *Note:* one nucleolar organizer suppresses its rival. × 2000 (from Babcock and Jenkins 1943).

The second kind of tetraploid arises less commonly but survives more commonly. It arises by doubling of the chromosomes in a hybrid between two diploid species. It might also perhaps arise from crossing two new auto-tetraploids of different species. It is known as an *allo*-tetraploid. The corresponding chromosomes of the two species, being dissimilar, pair rather with their own kind at meiosis in the allo-tetraploid. In consequence $2x$ pairs of chromosomes are formed; meiosis is regular; all the germ cells have $2x$ chromosomes, x from each diploid species. There is no segregation of dissimilar chromosomes and therefore no sterility. The new allo-tetraploid is fertile and true-breeding: it is a new species. This is the origin of *Raphano-Brassica* in experiment, *Spartina townsendii* in nature and *Aesculus carnea* in the garden, and of course of all polyploid species of plants grown for their seed like the cereals (Figs. 11 and 12).

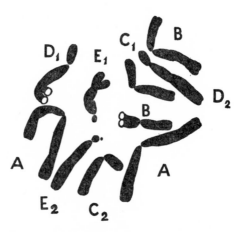

Fig. 12. The 10 chromosomes at mitosis in a pollen grain of *Paris quadrifolia* ($4x = 20$); the species is allo-tetraploid and three of the five homologous pairs are recognizably different. × 2000 (after Darlington 1941).

The mitotic chromosomes of allo-polyploids have one point of special interest. At their origin, each of the different haploid sets will have had a nucleolar organizer as a rule. But as we find them in nature there are fewer organizers than sets. In the allo-tetraploid species, *Paris quadrifolia*, we find that the second pair of organizers is no longer detectable. The reason for this is made clear by the behaviour of hybrids in *Crepis*. Here, in 1934, Navashin found that organizers from different species might suppress one another. They compete and in competing they have an order of dominance. *C. neglecta*, *C. dioscoridis* and *C. tectorum* lose their constrictions with *C. capillaris*, while *C. capillaris* loses its constriction with *C. parviflora* (Fig. 13). It is natural, therefore, that in an allotetraploid the organizer of one haploid set should suppress that of the other.

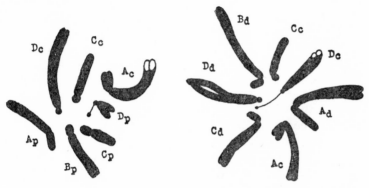

FIG. 13. Chromosome complements of two species crosses in *Crepis*. Left, *capillaris* × *parviflora*; right, *capillaris* × *dioscoridis*. The capital letters signify the members of sets, the subscripts the species from which they come. *Note:* in the first hybrid the separation of the satellite in the D chromosome of *capillaris* is lost, in the second, that in the D chromosome of *dioscoridis*, the result of competition between nucleolar organizers. × 3000 (from Navashin 1934).

Between the auto- and allo-polyploids there is of course every gradation since there is every gradation between intervarietal and inter-specific differences and between fertile and sterile hybrids. The situation may best be represented by saying that the fertility of a tetraploid is negatively correlated with that of the diploid from which it arose. This is true of polyploids arising in experiment. But when established allopolyploid species are examined we find that they have acquired special means of controlling meiosis which maintain fertility and give regular pairing although their chromosome sets are only slightly different. Thus in the 4x *Tulipa chrysantha* only bivalents are formed since there is only one chiasma for each bivalent. In $4x$ *Paris quadrifolia* the prophase begins at the centromere, all the chiasmata are localized near to the centromere and this does not allow quadrivalents to be formed. In $6x$ *Triticum aestivum* and in $8x$ *Dahlia variabilis* other special arrangements at prophase seem to have the same effect (see pp. 106, 207).

7. ODD AND EXTRA CHROMOSOMES

(i) Types

Following misdivision of the centromere, as we have seen, plants often arise with new chromosome numbers. There is a genuine change in basic number without any change in what the chromosome complement contains. But the fact that changed basic numbers so often come to distinguish species, genera, and still larger groups in the course of evolution shows that the change in content often follows. The breakage of a chromosome is often the first visible step in the breakage of a species.

Of quite a different character are those changes in chromosome number which do involve a change in the content of the chromosomes. In many species there are supernumerary chromosomes. There are 1, 2, or more extra chromosomes beyond the normal somatic complement. The basic number remains the same but a few odd chromosomes are added to it; and if several plants are examined they are found to vary in the number of these extra chromosomes.

The first type of such extra chromosomes consists simply of the ordinary members of the haploid complement represented a third time. The plant is partially polyploid. It is said to be *trisomic* (or even tetrasomic) for 1, 2 or more chromosomes. In most species, as we saw, such extra chromosomes cause unbalance. They produce abnormal growth; trisomics therefore do not compete in nature. It is this property which gives the basic number its constancy and hence its systematic importance. There are species, however, or perhaps even groups of species, in which the chromosomes have not developed such a differentiation. They have acquired or kept a similar balance among the brethren of one set. Relative to the rest each chromosome is, we may say, nearly neutral. Additions of extra chromosomes are nearly without effect. They are therefore occasionally and in some way advantageous.

Evidence of the nearly neutral chromosome is scarce and always demands careful study. It comes to us in two ways. On the one hand, in some species plants with extra whole chromosomes compete successfully with their normal relatives. This happens with *Narcissus bulbocodium* and *Clarkia elegans* in nature and with *Hyacinthus orientalis* (as we shall see in detail later) in cultivation.

On the other hand, in these species and in many others, for example, in *Trillium*, constrictions or the positions of blocks of heterochromatin give each chromosome a recognizable structure. It is then possible to see that pieces of the normal chromosome may be lost without adverse consequences. Plants are found with 1 chromosome smaller than its normal partner (Fig. 2).

Species therefore evidently vary in their fundamental chromosome organization or, as we may say, genetic structure. In some this structure is less rigid, more flexible, better buffered, than in others: and it is clear that where it is more flexible, where the chromosomes are more nearly neutral, changes in the basic number by gain, and less often by loss, of whole chromosomes willl more readily come about. Hence, in part no doubt, the variation in stability of basic numbers in different groups of plants.

(ii) Occurrence of B Chromosomes

The second type of extra chromosomes, at least in their extreme form, appear to be something altogether new. To them we give the name of B chromosomes. They occur widely in annual diploid flowering plants. They are distinguished from the A chromosomes of the ordinary complement by two properties. In the first place the B's, which are usually small, do not pair with the A's at meiosis but only with one another.

In the second place, B's vary in number among different individuals. They occur in both odd and even numbers and, what is more important still, in some plants of all populations in which they occur, they can be entirely dispensed with. They are thus not only supernumerary but what we may call subordinate or second-class chromosomes.

The B chromosomes must also be derived chromosomes: derived recently or ultimately from the A chromosomes. Since the A chromosomes vary between extremes of differentiation and neutrality, the B chromosomes are likely to vary also. In addition they are not likely to remain as they began. They will inevitably evolve, and they will do so in a pattern of their own. The B chromosomes we therefore expect to be very varied but none the less to have a character of their own (Fig. 14).

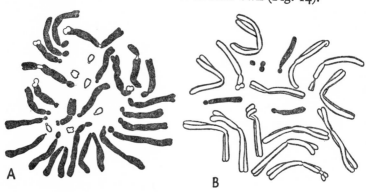

FIG. 14. Mitotic complements in the root showing B chromosomes. A. *Fritillaria imperialis:* $2n = 24 + 6BB$ (of the same size). B. *Secale cereale:* $2n = 14 + 5BB$ (of different sizes). *Note:* there are two pairs of V, ten of I in this species of *Fritillaria*. A. × 1400 (after Darlington 1939), B. × 2500 (G. J. Dowrick unpub.).

The fact that the B chromosomes can be dispensed with is connected with the fact that their presence cannot be detected in the external form of the plant. Unlike the different chromosomes of the A-type they have no specific action. They tend towards neutrality even in plants like maize in which the different A chromosomes are far from neutral. This used to be taken to mean that they were inert. Nowadays we do not draw this conclusion for we believe that the genes in these chromosomes are like the polygenes, which produce small and less specific effects and are the active basis of quantitative variation.

Not only are the B chromosomes active; they must also be useful. Not indiscriminately useful perhaps but useful in some special sense. There are various ways of showing this.

In any large natural or cultivated population of plants with B chromosomes we find that plants with and without B's exist side by side in competition (Table 3). In rye B chromosomes are found in populations from Afghanistan, Russia, Australia, Sweden, England, the United States and Japan. In the same countries other populations are found devoid of B chromosomes. The B's are usually of the same type indicating a common origin. Indeed, possibly the common origin may extend to the B chromosomes which are also found in several related wild diploid species of *Secale*.

TABLE 3

FREQUENCIES OF B CHROMOSOMES AMONG INDIVIDUAL PLANTS FROM (A) EXPERIMENTAL FAMILIES OF SEVERAL SPECIES, (B) LAND RACES OR CULTIVATED VARIETIES, AND (C) NATURAL POPULATIONS

Numbers of B Chromosomes	0	1	2	3	4	5	6	7	8	9	Total
Zea mays: 2x = 20 (D. and Upcott, '41)											
A { B × no B	109	58	3	—	—	—	—	—	—	—	170
{ 5B selfed	—	—	1	1	5	2	2	—	—	—	11
B { Golden Bantam	—	10	6	3							19
{ Black Mexican	—	2	6	8	12	5	4	4	1	1	19
Secale cereale: 2x = 14 (Müntzing, '50)											
A 2B × 2B	2	3	23	16	81	2	—	—	—		127
B { Turkish vars.	1055	9	50	1	1	—	—	—			1116
{ Afghan vars.	404	1	18	—	—	—	—	—			423
C *Poa alpina*[1]: 2x = 14 (Müntzing, '49)	—	—	3	2	11	4	9	1	4	—	34
Tradescantia paludosa: 2x = 12 (Riley, '44)											
C { W pop.	36	2	7	3	—	—	—				48
{ M & C pop.	45	15	20	11	1	—	—				91
Clarkia elegans: 2x = 14 (H. Lewis, '51)											
C Californian pops.	236	36	38	3	9	1	1	—	—	—	324

[1] B's lost in adventitious roots.

Now in rye and elsewhere the individuals, or races, or populations, without B's do not replace the individuals, or races, or populations, with B's. On the contrary the two states, with and without B's, are evidently in equilibrium, an equilibrium which changes in different races and populations but which may have continued for an indefinitely long time. The use of the B's therefore would seem to consist in their being advantageous to some but not necessarily to all members of a population.

The advantage of the B's moreover, depends on their numbers. If a few are advantageous it is quite certain that many are not. In *Sorghum purpureo-sericeum* a large number of B's lead to extra mitoses of the vegetative nucleus of the pollen grain. In rye they upset mitosis in both pollen and embryosacs. In both they thus reduce fertility and this is likely to be a general principle, since absolute neutrality of effect is not to be expected. Thus the B's are maintained in a species in spite of injuring fertility in the high doses in which they occur in some individuals. In the intermediate, which are no doubt the optimum, doses it is all the more clear that they must have a compensating value. Before pursuing this question we must first consider how B chromosomes arise.

(iii) Origin of B Chromosomes

How do the B's arise? We often see them newly arisen. But then they are most often perhaps of types which do not establish themselves. One example is in *Clarkia elegans*, where the chromosomes tend towards the low differentiation of *Hyacinthus*. They arise as supernumeraries through irregular meiosis in plants with rings of 4 or more chromosomes. Such plants are evidently hybrid for interchanges. All kinds of newly interchanged chromosomes will therefore be appearing some of which, small ones, will be suitable candidates for a B chromosome type of propagation.

Another mode of origin must be by fragmentation. In all plants from time to time centric fragments arise at meiosis and they may be passed on to the seedlings as supernumeraries. Usually they are the telocentric short arms of A chromosomes broken off by misdivision of the centromere. If they are nearly neutral or slightly beneficial they may be maintained in the race. In the orchid *Paphiopedilum wardii* there seem to be several kinds of B chromosomes which have arisen in this way and survive concurrently. B chromosomes arising by mis-division betray their origin by varying in number at mitosis in roots: after misdivision their centromeres are of insufficient strength for regular movement even in somatic mitosis.

In content of heterochromatin B chromosomes are most variable. They are sometimes devoid of it even when (as in *Paris polyphylla*) it is present in the A chromosomes: but they are often, as in maize and rye, highly heterochromatic. Perhaps they become more so in the course of evolution and thereby also become more susceptible to irregular crossing-over and hence to breakage and loss of end segments.

The extreme in divergence of B chromosomes from the behaviour of the ordinary complement is reached in *Parthenium*. Here the B chromosomes at meiosis begin to despiralize before metaphase. Such a departure from the regular cycle is nowhere else to be seen but in the sex chromosomes of animals. As with them doubtless it has an adaptive function for it dissolves the chiasmata and leaves all the B chromosomes unpaired at meiosis.

The products of misdivision and ordinary breakage have been found in B chromosomes of maize, rye and *Sorghum*. A change towards increased breakability is no doubt an early step in the differentiation from the more stable A chromosomes. Once all pairing and crossing-over with A's is cut off this breakable habit will quickly lead to the development of a special make-up. Its most notable characteristic will be a reduplication of parts: and the last stage of its evolution (which is not likely to be reached) would be the indefinite repetition of a single gene.

(iv) Adaptive Value of B Chromosomes

The physiological effects of B chromosomes can be measured in various ways. In maize as many as twenty B's have been found in one plant. They then add appreciably to the bulk of the chromosomes. They produce a kind of physiological poly-ploidy with the expected result that they increase the general size of cells: but it is in several ways a variable or elastic poly-ploidy. In the first place the B chromosomes, being usually shorter than the shortest A's and often present in odd numbers, do not pair regularly among themselves and are often lost at meiosis. This is true even in *Clarkia elegans* and *Crepis syriaca* where the B's are as long as the shortest A's from whose replication, as we shall see later, they may well have recently arisen.

In the second place, at somatic mitoses, as we have seen, B chromosomes are often liable to loss. Moreover, in certain species this liability is very exactly controlled. It gives the appearance of having been directed. They can be regularly perpetuated in the stem, which is virtually the germ-track of a plant, at the same time that they are equally regularly lost in the roots. This happens in two diploid grasses, *Sorghum purpureo-sericeum* and *Poa alpina*, and is analogous to a process of chromosome diminution known in the body cells, cells off the germ track, in many animals. It means that the species has two chromosome complements, one of which we might suppose was more suited to its reproductive processes and the other more suited to its vegetative life.

Have the B chromosomes a specialized value at one stage of the life cycle? Apparently there is evidence of such specialization in maize. That it is a very general effect is shown by the fact that a special mitotic device is found which piles up the B chromosomes in the pollen. In *Secale*, *Anthoxanthum* and no doubt elsewhere, the B's lag at anaphase in the first pollen grain mitosis. They pass undivided, that is double, to the generative nucleus leaving the vegetative nucleus B-free. In maize, and in *Sorghum* similarly, they lag at the second pollen mitosis, thus passing double to one of the sperm nuclei. This lagging in maize is due to a defect in the centromeres of the B chromosomes. The result, as Roman found by breeding experiments, is that the sperm with two B's goes to the embryo, that with no B goes to the endosperm.

In *Lilium* and *Trillium* the B's behave normally in the pollen. But in the embryo-sac mother cell they tend to pass to the embryo-sac and so to the egg cell nucleus. There are thus alternative routes through which the frequency of the B's can be boosted in the life cycle to compensate for the accidents of loss at mitosis or meiosis. These different mechanisms act to boost one or both of the nuclei which will accomplish fertilization; and their doubling before the stages where they are most needed will compensate in the life cycle for their losses at stages and in tissues where they are least needed.

A different kind of effect of B chromosomes has been noted by Fernandes in *Narcissus bulbocodium*. Here they seem to delay flowering. It may well be that quantitative effects of the B chromosomes will often have such specialized consequences which will favour the survival or propagation of the plants bearing them, and hence of the B chromosomes themselves. In this species of *Narcissus* Fernandes has also found chromosomes derived from interchange between A's and B's. These show us how B chromosomes, which have served a useful purpose by their unstable propagation, may be restored to the stable level of A chromosomes. In so doing they will raise the basic number of chromosomes in the species, as seems to have happened with the change from 7 to 11 in this very genus.

In all these things B chromosomes are doing what parts of the A chromosomes do in certain animals which undergo what is called chromosome diminution: but B chromosomes are found in animals as well as in plants. It is likely, therefore, that they achieve something more than can be done merely by the diminution of A's. What this may be is suggested by their distribution among species.

The habit of counting only single specimens of each species in the hope or belief thatchromosome numbers do not vary within species formerly tended to confirm this hope or belief and to conceal the existence of B chromosomes. If they seem to be more abundant among cultivated plants this is merely a symptom of the greater care given to the study of cultivated plants. We now know that hundreds of species possess B chromosomes, and we may surmise that a far larger proportion will later be found to have them.

When we compare polyploids with diploids we find that both (for example in the same section of *Tradescantia*) may possess B chromosomes, but in general they are less frequent in polyploids. Also they are so far unknown in the long-lived woody plants; and in maize, whose chromosomes have been more

POLLEN

EMBRYO
SAC

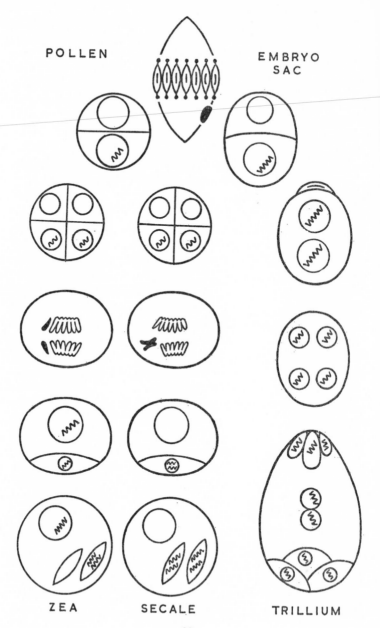

ZEA   SECALE   TRILLIUM

32

thoroughly studied than those of any other plant, B chromosomes are found in those aboriginal races in the United States most remote from the primitive home of maize in Mexico. In these the B chromosomes seem to have taken over the role of the heterochromatin in the A chromosomes. Such peripheral populations, as we shall see later, are the ones most subject to inbreeding.

These comparisons give us a clue to another aspect of the value of B chromosomes.

The A-chromosome point of view would be that in species where they occur, the first one or two B's are advantageous; higher numbers become disadvantageous. But since a constant number cannot be maintained there is merely a mean number which is higher or lower than the optimum according to the conditions of propagation at mitosis. This conventional view is not, however, the whole view. The B chromosomes need not be individually and inherently useful. Although they may be useful to every pollen grain, they need not be useful to every fertilized egg that comes to have them. Their value for the zygote may be not intrinsic but experimental. They may boost the variability and hence adaptability of a species. Such a boost will be less useful in a polyploid where it can usually be provided by gain or loss of whole chromosomes. It will be most useful in a diploid, especially a short-lived diploid which can afford to be experimental, or an inbred diploid which is forced to be experimental, or a diploid at the edge of its range.

FIG. 15. Three identified mechanisms of mitotic control adapted to the cell function and the population boosting of supernumerary or B chromosomes in plants. (i) Non-disjunction of chromatids and directed movement in the pollen grain to one male gamete at the expense of the other which effects preferential fertilization in *Zea*. (ii) Directed movement at the previous mitosis to the generative nucleus at the expense of the vegetative nucleus, in *Secale*. (iii) Directed movement to the embryo-sac forming dyad cell as opposed to the upper degenerating cell, in *Trillium*. By all these means the average gametic content of B chromosomes is raised above half of the parental content: a single unpaired B chromosome at first metaphase of meiosis appears once in all the egg cells, or twice in one or both generative nuclei of half the pollen grains. (Based on Roman 1950, Müntzing 1958, Darlington and Shaw 1960, Rutishauser, 1961.)

These are speculations which can be tested as we accumulate a knowledge of B chromosomes; provided we bear in mind that B chromosomes must have evolutionary cycles of their own. A chromosomes are all, in a sense, eternal. B chromosomes are either young or old, and we still know little of their evolution or of the different ways in which they may flourish or fade away or even return to the character of indispensable A chromosomes.

(v) Interpretation of Chromosome Numbers

For those studying the chromosome numbers of species the B chromosomes have a different kind of function: they have prepared a dangerous pitfall. They have created the outstanding exception to the principle that chromosome numbers are uniform for a species: for where there are B chromosomes the number is not constant for the seedlings of one family, or even, in some cases, for the individual plant. Varying in individuals, tissues and stages of the life cycle, they break almost all the rules of genetics and physiology which rest on the uniform behaviour of A chromosomes. Anyone surveying chromosome numbers by counting the chromosomes of one individual of each kind to which a distinct Linnean name happens to have been attached, will therefore go awry when he comes to a species with B chromosomes. It is to unidentified B chromosomes, therefore, that a great deal of unspecified variation in chromosome numbers within species recorded in the *Chromosome Atlas* must be due. To them also must be due a great many of the apparent discrepancies between the numbers recorded for single species by different authors.

In the study of plant systematics it is now clear that B chromosomes, instead of confusing the distinctions between species, can be used to classify and explain the variation and evolution within them.

8. CHROMOSOME SIZE AND DNA

The double helix of DNA makes the backbone and the main bulk of the chromosomes. Consequently the size of the chromosomes as seen under the microscope when stained is a measure of the mass of its DNA. This mass may also be measured by physical scanning of the preparation. It may then be seen to double towards the end of the resting stage in each nucleus; C becomes 2C when the chromosomes are replicated. Otherwise it is characteristic for all the nuclei of a plant (and indeed of a species) with the same chromosome number. This characteristic mass (C/2C) we find has a range among plant species of about 1 : 1000.

What is responsible for this great range of size? In the first place it is due to the chromosomes or chromatids being built round either one double helix or two, four, eight or more such helixes. Probably most fungi with very small chromosomes have one double helix. Most flowering plants, on the other hand, are *polynemic*. But plants with large chromosomes like conifers, Liliaceae, *Tradescantia* and some Ranunculaceae must be more highly polynemic than those with small chromosomes.

The other more obvious causes of size change are duplication and polyploidy. Later we shall see how polyploidy and polynemy interact in evolution.

II. PLANTS IN GROUPS

1. SPECIES AND SYSTEMATICS

SYSTEMATICS, or the grouping and naming of plants, has two purposes. The first is its practical purpose: it is concerned with the uses of plants for mankind in general, their uses for foods and fibres, drugs and oils, and last of all for ornament. The second is its fundamental purpose: it is concerned with their use for the advancement of knowledge, that is, for scientific research, for the study of the cells, individuals and populations of plants, of growth and heredity, of vegetation and evolution. All these things are important in relation to plants. But they are also important in relation to our knowledge of animals, man and the universe, indeed in relation to the whole body of human knowledge.

The first problem of systematics for both its practical and its fundamental purpose is that of species. These are the groups of plants to which names can be consistently and precisely given. They are also the groups of plants which can be defined consistently and precisely, if not permanently, by their behaviour. The test of the validity of species therefore depends on experiment: but there are also larger groups, genera, families and classes, in which plants are classified. These depend not on the behaviour of the plants and not on experiment, but on the convenience and conventions of the classifier.

To discover what names can be given precisely, being a question of experiment, the first serious definition of a species was necessarily an experimental one. It was given by John Ray in the introduction to his *Historia Plantarum* in 1686. This definition has proved to be sufficient down to the present day. Ray said that a species is a group of plants which breeds true from seed within its own limits (with *distincta propagatio ex semine*). Here we have a precise statement which is both empirical and analytic. It rests on fundamental genetic assumptions, assumptions which breeding experiments are continually

36

concerned to test. We know, in consequence of these tests, that the species, if breeding sexually, is a group whose members are inter-fertile and by interbreeding are thus able to exchange chromosomes and genes. In a strict genetic sense, therefore, they have their whole ancestry, and may have their whole posterity, in common. They are a genetic community or continuum.

The work of classifying could not, however, wait upon breeding experiments before reaching conclusions and assigning a name to each species. Indeed the mode of sexual reproduction of plants was not generally known when Linnaeus published his *Species Plantarum* in 1753. And the great systematist, making a virtue of necessity, felt free to exclude most of the consequences of sexual reproduction with the aphorism *de minimis non curat botanicus*.

The followers of Linnaeus have therefore from the beginning allowed themselves to take short cuts to classify plants by appearances. This means by inference from appearances. The more thorough investigators have taken the pains to infer from a uniformity of external form, or a continuity of geographical distribution, or both, that they were in fact dealing with a group of inter-fertile organisms, a breeding unit. This was a short cut through genetics. The task of dividing and naming the whole of the plant kingdom has, however, driven the less thorough investigator to take one short cut after another. He has often had to dispense, not only with *breeding* but even with *seeing*, the living plants. He has had to content himself with examining dry plants, aided only by a dry lens. This removes difficulties but it also loses opportunities: one is that of studying physiological and chemical variation; another is that of seeing the chromosomes and discovering how the species may have arisen—or indeed may now be arising!

In these puzzling and deceptive circumstances the systematist often formerly set his face, not only against new techniques, but also against the genetic theory which underlies them.

The first part of the genetic theory is the principle that standardization of the environment is necessary before differences of heredity can be inferred. The transplantation experiments of the genetic ecologists, using plants identified by their chromosome complements, have shown the unexpected variety of consequences arising from this principle.

The second part of the theory is that a surface comparison can give no evidence of the genetic relations of species. Not only can dominant conceal recessive, but a hybrid may be fertile and breed true either as a polyploid or in other ways that we shall describe. For this reason the product of self-fertilization may be, in a genetic sense, a hybrid arising from, and giving rise to, dissimilar gametes. If the dissimilarity is one of number a triploid is produced and it may be entirely sterile. But still this does not prove that it has arisen from crossing different species. For triploids frequently arise from the self-fertilization of fertile diploids: the parental gametes, not the parental zygotes, have been different. On the other hand, plants that appear to be entirely fertile may have arisen from crossing species or by triploidy: they may owe their seed to some form of non-sexual reproduction. In every group of flowering plants such *apomixis* occurs. It consists in the production of vegetative embryos which are purely maternal and compete with the sexual embryos if there are any.

The third part of the theory is that the unit of behaviour, of breeding, of evolution, is not an individual or a type considered alone. It is a mating group or population. New species may appear to have arisen by divergence from a type; but in terms of living processes they arise by splitting in the group, a splitting which is not at once externally visible. Changes in single genes on the one hand, or chromosome breakages or polyploidy on the other, often divide a species in two. They prevent crossing or they cause sterility of the F_1, the first generation cross. They thus establish a *genetic isolation* and this has the same effect directly as *geographic isolation* or a separation in space has indirectly, on the course of evolution.

All these principles we must shortly consider. The theory of systematics, on the other hand, has never been able to accommodate itself to the consequences of evolutionary change. It began as a theory of special creation. It continues to rest half-poised on the eighteenth-century fictions that the living world is inhabited by fixed species which exist, on the one hand, as 'types' represented by original specimens in museums and, on the other hand, as 'varieties' diverging from these types by some ineffable process of mutation. Sometimes the type names have been given to cultivated plants like *Tulipa gesneriana*, *Rosa chinensis* and *Primula sinensis*. And when the wild ancestors are discovered they have to be given subordinate varietal names. Sometimes the type names have been given to tetraploid forms as in *Lotus corniculatus* or *Coronilla glauca* whose 'varieties' are now known to be diploid.

The theory of systematics thus remains a theory of creation. The nomenclature and rules the systematist has based on this theory are unable to take account of evolutionary processes and relationships. The only difficulties he can notice are the flagrant ones where he is working with a welter of forms in groups where barriers to crossing are developed erratically or not at all. Lumping or splitting on any theory is then a matter of convenience. But on systematic theory the types chosen as the centres or foci of species, which are always arbitrary, are in these cases dangerously misleading. Intermediate forms between types are assumed to be hybrids. The term 'hybrid' is used not only for a first cross between dissimilar parents but also for any derivative of similar appearance. We know that there are different kinds of hybrid. We recognize under the microscope the true-breeding allo-polyploid and the true-breeding diploid hybrid which we shall discuss in *Oenothera*. We can distinguish two kinds of triploid, both hybrids, but both of them resulting from self-fertilization: those which attempt to be sexual but fail to form seed; and those apomictic plants which form fertile asexual seed.

But sterility to the systematist has usually been of one kind, the kind which may be used to confirm hybridity. Of course, this is true sometimes; but one cannot tell without seeing the chromosomes, using the pollen, or growing the seed.

Over and above these fictions, errors and half-truths, which can be remedied by a study of the chromosomes, are certain rules of procedure in systematics which are beyond the remedy of individual effort. By these rules the names of species are taken, not from the author who has understood the species *best*, but from the one who has described it *first*, and probably therefore knew least about it. This principle has had the effect of turning attention away from observation and experiment and towards quotation and repetition and the pursuit of forgotten authors. It has also had the effect of stimulating changes of names by museum workers.

Chopping and changing for their own sake have brought systematic botany (and zoology) into contempt. Their effects have long ago been deplored by great naturalists such as Gilbert White and Joseph Hooker. The least of these effects has been to lead practical men who are seriously concerned with growing plants, breeding plants, and using plants, to keep the old name[1] or to prefer the certainty and fixity of the popular name. This is most obvious where systematists have tried to introduce a system of types and varieties for the classification of rapidly evolving cultivated plants like the species of *Brassica* where no practical man has ever paid attention to names which (since they ignore chromosome numbers and breeding behaviour) afford no indication of past ancestry or future progeny.

Now it is not and never will be practicable to carry out experimental tests for classifying, species by species, the whole of the flowering plants. We can, however, apply experimentally verified theories. And we can apply them at once, with the help of experimentally verified methods, time-saving short cuts, to a sample of flowering plants. The one important short cut is the use of chromosome studies for they reveal at once three important properties, the patterns of variation in genera, the modes of origin of species, and the kinds of reproduction of individuals.

We can now turn to the question of how we are to graft the Linnean system on to the chromosome system or vice versa: it is the great question for both the practical and the fundamental aspect of classification. Can we, for example, assign chromosome complements and numbers to each Linnean species and its varieties, attaching the names of the authority

[1] As with *Pyrus japonica* (of Sims, 1803), which is still so known to gardeners although this name was botanically superseded by *Chaenomeles lagenaria* of Koidzumi in 1909.

to each? Can we add the chromosomes as an appendix to the otherwise sound work of museum classification? The answer is that we cannot.

We must make clear at once that the two systems are unavoidably in conflict. There can be no compromise between them. Either we can follow a scholastic system which pretends to finality in details as well as in principles, or we can follow an experimental system which in details attempts only to be provisional, reserving dogmatic assertion for matters of principle.

The chromosomes of the actual type specimen cannot be known for certain. To pretend that we know them would be to impose a guess, often a false guess, and always by implication, a fraudulent guess, on the reader. The chromosome number is itself a definition of a plant—the unique plant whose chromosomes were examined. The chromosomes of a living plant themselves often constitute a definition of greater significance than any character of a dead plant. When they can be characterized by a number or a diagram in respect of which related groups differ, then the number or diagram may be more useful in defining the species than the name of the author responsible for a description of a dead plant. And the number or diagram which we take for our authority should be, not the first, but the last to be published; or the one with the most information on other relevant matters such as chromosome behaviour at meiosis, mode of reproduction, sexual fertility, and so on.

How this system works has been shown by studies of *Crepis*. By comparing the chromosomes Babcock was able to throw many species, like *Youngia tenuifolia*, out of the genus *Crepis*. He was also able to take what had been assigned to five other genera and fit them into their proper places in *Crepis*. Thus:

> *Phaecasium pulchrum* became *Crepis pulchra*.
> *Cymboseris palaestina* became *Crepis palaestina*.
> *Pterotheca sancta* became *Crepis sancta*.
> *Rodigia commutata* became *Crepis foetida commutata*.
> *Zacyntha verrucosa* became *Crepis zacyntha*.

Such changes can be quickly vindicated by chromosome

studies (Fig. 16). But these are only the first step to a genetic understanding. Only later could Babcock show that the old genus *Rodigia* differed merely by three genes segregating in a Mendelian manner from the typical *Crepis foetida*.

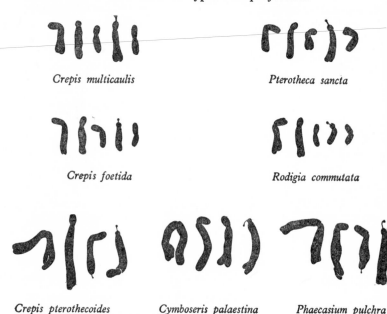

Crepis multicaulis

Pterotheca sancta

Crepis foetida

Rodigia commutata

Crepis pterothecoides   Cymboseris palaestina   Phaecasium pulchra

FIG. 16. Chromosome sets of species assigned to five other genera together with those of closely related species in *Crepis* with which they have now been assimilated. × 2000 (from Babcock and Jenkins 1943).

Another example from *Haplopappus* is worth illustrating. Some 200 species of this genus of the Aster tribe are distinguished in western N. America. Within the range of what was known as *H. gracilis* the chromosomes revealed the existence of two discontinuous groups (i) *H. gracilis* s.s. with 2 chromosomes in its haploid set and up to five small B's in some populations; and (ii) *H. ravenii* with four haploid chromosomes. These two genetic species can be crossed experimentally but the hybrid is infertile and the cross is evidently not effective in nature.

It seems likely that in the origin of *gracilis* from *ravenii* the B chromosomes are the dispensable smaller products of unequal interchanges whose loss gives the reduction of chromosome number.

In the Linnean past the confidence with which we have used botanical names has often been a measure of our ignorance. This was true of the hundred tuberous *Solanum* 'species' in the first edition of the *Chromosome Atlas*. When we know such groups of plants a little better we hesitate to assign them to species. Only when we know them extremely well, only in those genera like Gossypium, Nicotiana and Triticum where a prodigious effort, experimental as well as descriptive, has been concentrated on a limited field, can we begin to speak with conviction based on knowledge; and such a happy issue has been reached only, be it noted, where great economic importance has attracted attention to the group. Otherwise we have to use our names with diffidence.

2. THE IDEAL MODE OF REPRODUCTION

If species are groups which breed true from seed it behoves us to find out what happens when a plant bears seed. Chromosome studies and breeding experiments have shown us that in fact more diverse processes are hidden within the formation of seed than Linnaeus or Darwin, could have dreamt to exist.

Take first the simplest situation. There is the species which is entirely diploid. It is cross-breeding, sometimes owing to some externally visible mechanism, but more often owing to some genetic incompatibility system, which favours cross-fertilization. Its meiosis is regular and crossing-over, as shown by chiasmata, is distributed throughout its chromosomes. Owing to the habit of crossing, all individuals are moderately heterozygous for gene and structural differences between their homologous chromosomes; and owing to the habit of crossing-over and segregation at meiosis these genetic differences are recombined at each fertilization. Each plant is a unique individual but it resembles its sibs closely and other members of the group in proportion to the closeness of its relationship by descent.

What is this group? It is a body of individuals within which there is or can be a free exchange of genes. It extends geographically in space and historically in time. It constitutes a continuum within which no sharp breaks, or at least no sterilizing breaks, occur. It may constitute the whole of what the systematist calls a species but it sometimes constitutes a very small part of a species.

Without qualification this ideal system can probably be attributed to no large group of plants; but it can be applied to many small groups, and it is these groups which, in a sense, possess the future. They are the ideal ancestors. It is they, and they alone, which can give rise to all the other kinds of genetic system and all the other kinds of species. They way they do so is always by restricting genetic recombination, a process of degradation, usually irreversible, of the ideal genetic system.

The simplest of such changes may well be a restriction in the distribution of chiasmata and crossing-over since differences in this respect distinguish species wherever suitable observations can be made. For example, *Allium cepa* has distributed chiasmata; *A. fistulosum* has chiasmata localized near the centromeres. Within species the process of change has been reversed experimentally. In outbred rye the chiasmata are localized near the ends. Inbred strains have been raised by Rees in which they are evenly distributed. The effects of this simple change from outbreeding to inbreeding have been studied in some detail and they are important because, as we shall see later, the change is characteristic of the domestication of wild plants.

Inbreeding overtakes our ideally (or primitively, to use the morphologist's expression) outbreeding stock in many ways. A stock may lose its incompatibility gene; or, by becoming tetraploid, it may become compatible and self-fertile. It may acquire a gene for cleistogamy. It may, in spreading over a larger territory, outrun its insect pollinator. Whatever the cause of this fateful change, every inbreeding individual is thereby likely to become the parent, by the same stroke, of a pure line, and also of a genetically isolated unit. Gene exchange ceases. Instead of a single net of descent there emerge separate lines

44

of descent. All genetic differences are fixed in these lines and each pure line is a genetic species in the sense of Ray. We must now consider how this happens and how, in certain circumstances it is frustrated.

3. OUTBREEDING, POLYMORPHISM AND SEX

If outbreeding is the ideal mode of reproduction, how do plants avoid inbreeding? The bulk of flowering plants have the organs of both sexes in the same flower: they are said to be hermaphrodite. This situation favours self-fertilization; the reproductive history of the flowering plants is therefore largely an account of different ways in which they evade self-fertilization or escape from its consequences.

The commonest but feeblest means of escape is by using some device of structure or development which prevents the pollen of the hermaphrodite flower from falling on its own stigma. Such a device, however, does not prevent the pollen falling on another flower of the same plant which (as Darwin and Mendel found) still constitutes self-fertilization. Inbreeding can be effectively prevented only by the occurrence of genetic differences in the population which, first, distinguish between one plant and another and, secondly, favour crossing at the expense of selfing.

Such differences occur in every family and in most species of flowering plants. They belong to three main systems, incompatibility, heterostyly and dioecy. All of these are subject to medelian inheritance. All are therefore due to differences between chromosomes although only in the third system, dioecism, have the differences so far been seen under the microscope.

Incompatibility. In the first and most widespread system there are differences between the plants, and between the pollen grains they produce, in respect of certain incompatibility genes. Pollen grains carrying one allele will grow only on styles of plants whose two alleles are both of them different. For example, a plant with genes which we may label S_1S_2 produces pollen grains S_1 and S_2 neither of which can grow on its own styles though both can grow on another plant's styles of the constitution S_3S_4. The S_1 pollen will also grow on a style S_2S_3 and S_2 on a style S_1S_3.

TABLE 4

TYPES OF PROGENY FROM INBREEDING AND OUTBREEDING IN AN
INCOMPATIBILITY SYSTEM

Styles	Pollen Grains		
	$S_1 + S_2$	$S_2 + S_3$	$S_3 + S_4$
S_1S_2	nil	S_1S_3, S_2S_3	$S_1S_3, S_1S_4, S_2S_3, S_2S_4$
S_2S_3	S_1S_2, S_1S_3	nil	S_2S_4, S_3S_4
S_3S_4	$S_1S_3, S_1S_4, S_2S_3, S_2S_4$	S_2S_3, S_2S_4	nil

Note: (i) Crossing between identical types of plant, like selfing, leads to nothing.
    (ii) Reciprocal crosses often differ in the progeny resulting.
    (iii) Remote crosses give more success and more diversity than close crosses.

To put this situation at once in its physiological and its evolutionary perspective we have to say that the style selects or sieves the pollen tubes which it will permit to grow in its tissue. It does so by a chemical selection of those whose gene products are different from its own.

It will be seen that in this system every individual produced by fertilization will be heterozygous in respect of the incompatibility gene. It need not be heterozygous for other genes even on the same chromosome if they are capable of separating themselves from the incompatibility gene by crossing-over. But close to the critical gene there will be a zone of permanent heterozygosity. Within this zone other gene differences have developed assisting the main gene in its chemical activity. Further, in a species following this system, there may be hundreds of different plants of each incompatibility type which therefore constitute mutually incompatible groups.

46

Heterostyly. In the second system the population usually contains only two types or groups of plant in regard to incompatibility. These differ in the structure of their flowers. There is a long-style short-stamen type, known as the *pin*, and a short-style long-stamen type known as the *thrum*. Not only is pollination by insects physically favoured between, as opposed to within, these two types but the development of this physical difference is found to be secondary. It has been preceded in evolution by a physiological difference, like that working through the S gene, whereby the cross-pollen, pin on thrum or vice versa, grows better than the self-pollen, pin on pin or thrum on thrum.

Further, combined with the changes in growth are changes in cell structure. The pin pollen grains are smaller than the thrum. And alighting on the thrum stigma they find shorter papillae than on the pin stigma of their own flowers. Thus there are a whole cluster of differences which have arisen by mutations within a gene complex to give a useful and effective combination of properties.

This system, known in hundreds of genera, is familiar to us as *heterostyly*. It was noticed in *Primula* by Linnaeus but was studied later by Darwin and others. It is due to the existence in species of two types, one homozygous, the other heterozygous. These two types being compelled to cross with one another maintain an equilibrium with equal numbers in succeeding generations. The result has the form of a mendelian back-cross. The thrum pollen, it will be noticed, must be of two kinds carrying opposite alleles. But it all has the thrum properties of growth and form which are regularly imposed on it by the mother plant. And the thrum type is permanently heterozygous for the S-s alleles while the pin type is always homozygous for this complex or *super-gene* (Fig. 17), in the chromosome. No differences of size have however yet been noticed between pin and thrum chromosomes.

Dioecy. The principle by which a species is divided into individuals of two kinds which have to cross and are produced in about equal numbers, is the basis of the third outbreeding system. This is the same system of sexual differentiation so

E

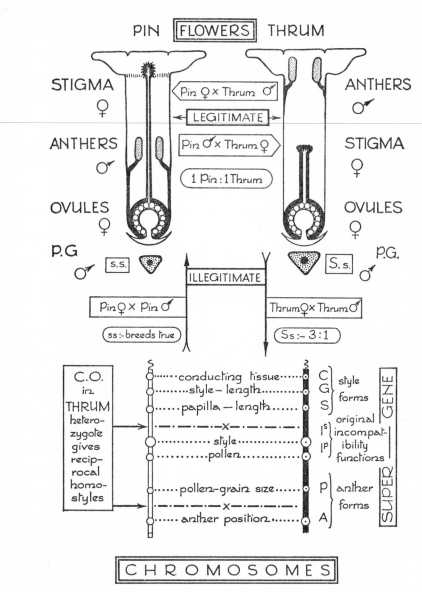

FIG. 17. The basis of balanced polymorphism in the form of heterostyly in the chromosomes of a species of *Primula* (from Darlington 1971).

widespread in animals. Here there are male plants producing only stamens and pollen and female plants producing only ovules and seeds. We might expect again, as in heterostyly which is about equally common in plants, that one form, one sex, would be heterozygous, the other homozygous. No species with very large chromosomes is dioecious. But wherever they are large enough to be studied the chromosomes reveal the expected difference. The male is usually the heterozygous sex and it is seen to be distinguished by a pair of chromosomes of unequal size. These are labelled X and Y and are said to be sex chromosomes as opposed to the normal pairs which are known as autosomes. The relations of the sexes being reciprocal in reproduction, their proportions in the population are equal as we may see diagrammatically.

The principle is the same as that with heterostyly. It is again a back-cross. And again other differences have attached themselves to the original gene difference and some of these have involved changes in size of the chromosomes. It is in this way that the sex chromosomes have become distinguishable from one another.

Sexual differentiation has lasted in animals, we may suppose, without a break through the evolution of groups as large as the insects and the vertebrates. The corresponding system in flowering plants by contrast has arisen here and there and has never persisted so as to embrace large systematic groups (Table 5). Moreover it is apt to be submerged in polyploid species, surviving in only a few. In these respects again dioecy in plants resembles heterostyly.

In their forms and relations the sex chromosomes of plants show certain universal properties and also certain variations which together tell us how their machinery works.

In the homozygous sex the two X's pair as a rule by a chiasma in each arm. But in the heterozygous sex X and Y pair as a rule by a single chiasma, usually terminal, in one arm; it is always the same arm. Thus X and Y are evidently the same in one segment, the homologous or *pairing segment*. They differ in the rest of the chromosome, the non-homologous or *differential segment*, in which no pairing, no crossing-over, no chiasma formation, occur. The differential segments of X and Y are therefore separated in evolution and can diverge in evolution as though they were chromosomes in different species.

TABLE 5

FREQUENCY OF GENERA WITH SPECIES HAVING DIFFERENT FORMS OF SEXUAL DIFFERENTIATION IN BRITISH FLOWERING PLANTS (after Lewis 1942)

Genera	hermaphrodite	monoecious	dioecious
Species all one type	468[1]	28	15
+ dioecious species	9	2	—
+ monoecious species[2]	1	—	—

[1] Including those with species having populations with a proportion of female (i.e. male-sterile) plants in equilibrium. Such species are said to be gyno-dioecious.

[2] Having male and female flowers separate on the same plant.

Thus we see why it is that X and Y chromosomes have themselves changed in size. The differential segments have diverged. Usually as in *Humulus lupulus* X is larger than Y; sometimes as in species of *Melandrium* Y is larger than X.

This divergence can take many striking forms of which we may give one example. In some species, notably *Humulus japonicus* and in all species of *Rumex* there are two short Y chromosomes in the male instead of one larger one: XX♀ and XY_1Y_2♂. The Y chromosomes pair by single chiasmata with opposite ends of the X. In these cases clearly the X has two pairing segments at opposite ends with a single differential segment, which includes its centromere, between them. We suppose that there was an original Y chromosome which split by misdivision at the centromere within its differential segment at an early stage of its divergence from X. Thus two telocentric half-chromosomes would arise which later acquired short arms (Fig. 18).

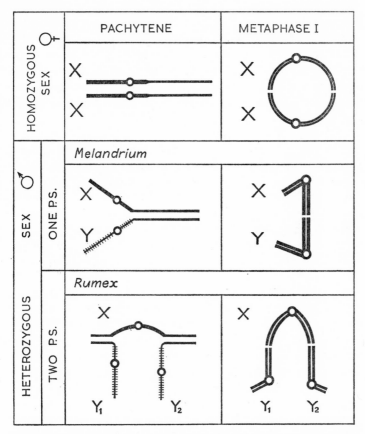

FIG. 18. Diagram showing relations of the sex chromosomes in plants with the best understood types: those with one and with two pairing segments. The two stages are: (i) Pachytene with complete pairing of the homologous or pairing segments, and of the differential segments also in the homozygous sex. (ii) First metaphase with formation of one or two chiasmata which, moving to the ends, have become terminal. The examples are both with the male as the heterozygous sex. Circles are centromeres; thin lines, pairing segments; thick lines differential segments of X; and crossed lines of Y. In *Melandrium* Y is larger than X; in *Rumex* there are two different Y's (cf. Westergaard 1958, Darlington 1958).

The early stage of the *Rumex* system, before the Y has split, no doubt occurs frequently but has been seen only in animals.

A cluster of minor, or indeed minute, adaptations have, as we saw, attached themselves to the diverging chromosomes in the development of incompatibility and heterostyly. The same process has occurred in the evolution of the sex-determining and sex-differentiating system as the chromosomes have diverged. There is thus not a single gene difference between X and Y but a complex of genes or a super-gene which works as what we may call a micro-adaptive unit.

One of the first examples of this was noted very early in *Melandrium*. Correns found that sparse pollination always gave a $1 : 1$ ratio of the two sexes in the progeny. But copious pollination yielded an excess of females. He assumed that the pollen tubes which are male-determining (with a Y chromosome) grew more slowly down the style than those which are female-determining (with an X). It thus brought the male plants up to equality only when the pollination was sparse enough for every pollen tube to find an ovule. In this way we can see that evolution of the genes making the X-Y difference in a species is adjusted to give the most fruitful proportions as well as the most effective relations of the two sexes.

They establish polymorphisms in the species which are balanced in the sense of Ford. That is to say their advantage consists in the diversity that they create because each alternative increases the value of the other. But these reproductive polymorphisms are remarkable in another way. First they constitute diversity in themselves. Secondly each alternative is not only of value to the other: it is necessary to the other for the reproduction of the species. Thirdly, the reproduction that they allow is such as to encourage general hybridization and hence diversity in all other respects.

A number of different evolutionary sequences follow from these connections. The condition of hybridity is important in all of them. Take the Y chromosome and the thrum supergene. They have never existed as homozygotes since they started their present career millions of generations ago. The same is true of the incompatibility alleles. All of them are perpetually hybrid. Thus systems which began as means of ensuring genetic recombination have become by their evolution the means of enforcing a localized hybridity which itself in a different way encourages hybridization and recombination. Hybridity and recombination become, as it were, a common pursuit. We now have to see how their alliance may break down.

4. INTERNAL HYBRIDITY

When a change towards inbreeding takes place in a group of plants it sometimes fails to produce homozygous lines. The homozygotes do not live; or at least they do not survive and reproduce in competition with the heterozygotes. Only the heterozygotes are free from any lethal or deleterious effect and they therefore breed true, thus:

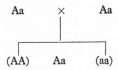

Very often, side by side in the same chromosomes with such self-preserving gene differences, are structural differences. The hybrid chromosome then also breeds true for its structural difference. The allele *a* may lie in a segment inverted relative to that in which *A* lies. The plant is then an inversion hybrid, and it may even breed true to being a hybrid. If the inversion is short there will be no crossing within it. The gene and the inversion will then make a single difference inherited in one piece for ever: the two will be able to make a super-gene.

Again, a chromosome may have broken in two at the centromere. Thus, if the two arms are labelled C and D we find 3 chromosomes CD, C and D in the complement. This hybrid condition is found not only experimentally but also in some populations near Leningrad of the cross-breeding species *Campanula persicifolia*; and in *Spiraea filipendula* in England it is found in every individual. The species has become a true-breeding hybrid in respect of a structural change in the chromosomes and also (we must suppose) in certain genes in those chromosomes (Fig. 19). The result is parallel with XY_1Y_2 in *Rumex*.

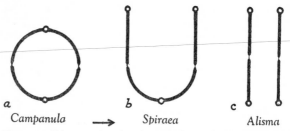

a *Campanula* ⟶ b *Spiraea* c *Alisma*

FIG. 19.   Diagram showing, at meiosis in the hybrid, b, and in the two homozygotes a and c, the relations of telocentrics with the chromosomes from which they arise. In *Campanula* a a and b are found; in *Alisma* a and c; in *Spiraea* apparently only b. (See fig. 5.)

This process has overtaken whole sections of the genus *Oenothera*. The original cross-breeding species have turned towards inbreeding. In doing so they have preserved their hybridity which happens to be marked by interchanges of segments between chromosomes: AB and CD from one gamete form a ring at meiosis, with BC and DA from the other. The interchange homozygotes are also the gene homozygotes since crossing-over and chiasmata are localized near the ends of the chromosomes. These homozygotes die. The hybridity of those that survive (and constitute the species) we can detect by crossing the different species which give in first generation, the F_1, four kinds of hybrids instead of one kind. We can also detect it by observing the chromosomes associated in rings at meiosis as a result of the interchanges of segments which have taken place between them in the remote past. Let us see how this comes about.

In all plants and animals from time to time chromosomes interchange segments as a result of accidental breakage. Thus two old chromosomes, which we may label by their two arms, AB and CD, give two new chromosomes, BC and DA. In this way an interchange hybrid is created whose chromosomes associate at meiosis to give a ring of four, thus:

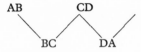

Chromosomes next to one another in the ring go to opposite poles in the mother cell so that two kinds of gametes are formed AB + CD and BC + DA. After self-fertilization these units, as in a Mendelian hybrid, give a ratio of 1 : 2 : 1 of seedlings which are

1: pure	AB — AB + CD — CD	(2 pairs)
2: hybrid	AB — BC — CD — DA	(ring of 4)
1: pure	BC — BC + DA — DA	(2 pairs)

Such plants have been found in nature in *Campanula persicifolia*, *Clarkia elegans* and dozens of other species. If they are self-fertilized the pure types are frequently weaker and are eliminated. Natural selection helps the structural hybrid which thus breeds true.

This selection in favour of hybridity is strongest if the plant's breeding opportunities are forcing it in the opposite direction, that is, towards inbreeding. Since enforced inbreeding, as we shall later see, is characteristic of modern plant breeding, this change in the breeding system may be held responsible for the appearance of strains in garden races of *Matthiola incana* with rings of 4 or 6 chromosomes.

How this comes about has been shown experimentally in *Campanula*. By crossing plants with different rings of 4, i.e. with different interchanges, rings of 6, 8, 10 and 12 chromosomes have been successively built up. When any of these ring-plants is crossed with a similar ring-plant smaller rings and simple-pairing plants appear in the progeny. Such interchange-homozygotes are not too homozygous in respect of their genes to survive; but when the ring-plants are self-fertilized any progeny with simple pairs are as near as possible to being absolute homozygotes. They therefore die and the rings breed true: they breed true to being hybrid.

Ring-formation is therefore a visible marker in the cell of the hybridity of a plant. It is also an easy answer for a species to any circumstances which force inbreeding upon it. It is not surprising, therefore, that ring-formation has developed in a number of genera, one interchange following another, up to the point at which all the chromosomes are included in the ring as follows:

$$10 \quad \text{in} \begin{cases} \textit{Paeonia californica, } \text{N. America} \\ \textit{Chelidonium majus, Europe}[1] \end{cases}$$

$$12 \quad \text{in} \quad \textit{Rhoeo discolor, } \text{C. America}$$

$$14 \quad \text{in} \begin{cases} \textit{Oenothera } \text{spp., N. and S. America} \\ \textit{Isotoma petraea, } \text{Australia} \end{cases}$$

In *Rhoeo*, and *Chelidonium* only the first and last stages of this evolutionary process are known, but in *Paeonia, Isotoma* and above all in *Oenothera*, where it has happened independently in three sections of the genus, the successive steps are still to be seen (Fig. 20).

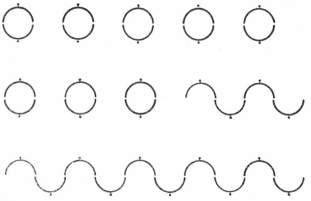

FIG. 20. Diagram showing the transition in the development from five bivalents to a ring, or chain, of 10 chromosomes.

The largest rings in these species it will be noticed are those which absorb 14 chromosomes. This does not mean that species with larger chromosome numbers do not begin to develop rings. Any diploid species with median centromeres and regular terminalization of chiasmata is likely to have interchanges floating in its populations which give rings of 4 or even 6. A

[1] This is a plant from Sapporo Botanic Garden which must, however, represent a European race. The wild English form has 12 chromosomes regularly forming six pairs (Nagao & Masima 1943).

few species with 16 or 18 chromosomes begin to form larger rings but fail to attain a complete ring probably because they run into mechanical difficulties. We shall consider one example later in *Clarkia*, a genus related to *Oenothera*.

The development of ring formation has striking consequences for the external appearance of a group of species. In the intermediate stages each new species becomes highly heterozygous and also somewhat unstable. It is liable to produce 'mutants' on a large scale. Such indeed were the mutants of *Oenothera* discovered by de Vries. Some of these were connected with evolutionary change as he thought; but they were connected as consequences of a special process rather than causes of a general process: for most of them were merely trisomics (with one extra chromosome), triploids and tetraploids, all of them, even on a short view, quite off the track of evolutionary advance. On a long view, on the other hand, since their hybridity is maintained only by cutting out sexual recombination between pairs of chromosomes at meiosis, we know that the whole system must be an evolutionary blind alley.

At the last stage in *Oenothera* the ring of 14 has become stable. The species is a 'complex heterozygote'. It forms only two types of gamete which survive only by reproducing the parental type of zygote and their difference can therefore be discovered, as Renner discovered it, only by crossing species.

At the last stage of all it is possible for only one species to be left. This is the case with *Rhoeo discolor* to whose history we can now add a footnote. Side-by-side with the purple-leaved complex hybrid ring-of-twelve species in British Honduras Wimber discovered less robust green plants forming six bivalents. When he crossed the two together they gave twin progeny of the two chromosome types like the two parents, hybrid and non-hybrid:

$$(12) \times 6(2) = (12) + 6(2).$$

This result indicates that the green form in nature is not the ancestral species but is the result of crossing-over of the lethal genes between the two complexes (as happens in *Oenothera*) thus allowing homozygous forms of one complex to survive.

5. AUTO-POLYPLOIDY AND ASEXUAL REPRODUCTION

The production of an allo-tetraploid in nature by doubling the chromosomes of a diploid hybrid or by the crossing of two autotetraploids creates a species, often a new species. It does not arise from a change in the breeding system, on the contrary by its origin it changes its breeding system: it is now (if it survives) the whole of a species with a population of one. It becomes in one stroke a self-breeder or no breeder at all. The new polyploid must therefore lose its system of incompatibility or of sexual differentiation. It has, however, two profitable and not incompatible evolutionary courses open to it. It can, by the formation of quadrivalents and by segregation of chromosomes from its two diploid parents, by a kind of secondary segregation, produce new variation which is likely to be of adaptive value. It can, for example, use this variation in adapting itself to reduce chiasma frequency, to avoid the formation of quadrivalents and to acquire the character of a good functional diploid; this result is achieved almost perfectly by the polyploid cereals. For them, of course, true-breeding sexual fertility is the highest need, and we shall see later exactly how they achieve it.

Quite otherwise is the case with the autopolyploid forms arising in nature. In view of the reduced fertility of autotetraploids it is surprising that autotetraploid races and species are by no means uncommon in nature. For this there are probably several reasons. Irregular meiosis with the formation of quadrivalents gives them a new kind of variability which must help them in colonizing new territory. The giant form is often advantageous for the same reason: size can always be reduced by genetic unbalance but it cannot so readily be increased. Again the fertility of plants is usually much in excess of what they need for their propagation. Especially is this true if they can rely on vegetative propagation to keep them going. Indeed, we may say that a margin of superfluous fertility is apt to be used by the species as the working capital for evolutionary experiment. And the polyploid itself is just such an evolutionary experiment.

Just as most diploid plants include some polyploid cells, so most diploid species include some polyploid individuals. In many species, like *Ranunculus ficaria* in England or *Narcissus bulbocodium* in Spain or Portugal, tetraploid and even hexaploid forms are probably arising all the time and largely depend on vegetative propagation to perpetuate themselves and to exploit a local ecological situation or to migrate into a new geographical area.

The existence of triploid varieties and species which we find here and there in any list of chromosome numbers allows us to carry this argument further.

In most diploid species of plants one seedling in every few hundred is a triploid. This triploid is more vigorous than its parents and brethren. It is also sexually sterile and not therefore burdened with the expense of pollen or seed production. If it has the capacity to multiply vegetatively, by tubers or runners or viviparous flowers, this capacity will often enable it to compete with its diploid progenitors. It may live side by side with them, as we find it doing in hundreds of bulbous and tuberous species, or it may even displace them. In most of these, as for example in the tuberous *Solanum* group, the diploid still survives, but in many, as in *Tulipa* and *Fritillaria*, it has been displaced. Very often the chromosome number has been our first indication that the species was relying solely on these reproductive expedients.

Vegetative propagation destroys the gene exchange on which our ideal species depends for its variability and also its continuity. No change is possible except that of sudden gene and chromosome mutations, which are irreversible. It is therefore a stereotyping system. It brings evolution to an abrupt end.

Sterile triploid species like *Tulipa saxatilis* have, we may say, made their last evolutionary experiment. They are cast in an invariable mould. Like *Rhoeo discolor*, they are species in which we expect no progress. If the environment to which they are adapted seriously changes, they must disappear: but for the short term situations, and above all, as we shall discover, for the uses of man in cultivation, assisted by grafting, these shortcomings may be overlooked.

Another means of suppressing sexual reproduction arises when meiosis and fertilization fail together and in co-ordination so that diploid may beget diploid, or triploid beget triploid. This new method of reproduction, which we recognize as apomixis, has been supposed to occur here and there among flowering plants. It has also generally been supposed to avoid all segregation and recombination and thus to be completely asexual in effect. Both these views require correction. In the first place, apomixis occurs very widely. We may now suspect that it occurs in some form or degree in most families if not in most genera of flowering plants. It therefore deeply affects the practice as well as the principles of the classification and naming of species. In the second place, it occurs in many gradations. How this happens we must consider.

TABLE 6

DIAGRAM SHOWING EVOLUTIONARY AND USUALLY IRREVERSIBLE CHANGES IN THE BREEDING SYSTEM WHICH ARE ASSOCIATED WITH THE ORIGIN OF SPECIES

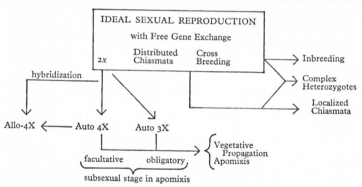

The replacement of sexual reproduction by apomixis can befall only plants which somehow or in some degree have lost their sexual fertility. By mutation, by hybridization, in either a first or a derived generation, or by triploidy, sexual reproduction may come to fail. If it fails, as happens with most triploids, the plant may simply not set seed. The plant may continue vegetatively, but that is the end of its reproductive history. If meiosis fails, however, and an unreduced egg or a purely

60

vegetative bud in the ovule can develop without fertilization, the race or stock will have a new lease of apomictic life: but both the suppression of sexual reproduction and its replacement may be incomplete. Then apomixis is facultative or optional or, as we may say, reproduction is versatile.

Versatile reproduction is best known in *Rubus* and *Poa*. By its means a great range of haploids and triploids and sexual and asexual diploids are sometimes produced within the same fruit. Their proportions, the result of competition among different possible types of embryos, depend on whether the flower has been self- or cross-pollinated.

In *Rubus* the competition is between reduced and unreduced embryo sacs which are fertilized or unfertilized. In *Casuarina* however it is also between fertilized egg cells and unfertilized synergids. But in either case the result is to produce mixed populations with different degrees of polyploidy.

Apomixis may, however, be complete and obligatory. Indeed in a triploid it nearly always must be so. It then establishes in each line in which it arises a species with a character of its own; for every line is then as well isolated from every other as though an ocean stood between them: but even there the result is often of a less uniform character than we should expect of a purely vegetative line. The breakdown of uniformity comes about by what we describe as *subsexual* reproduction, and this happens in several ways of which two are probably widespread.

The first mitigation of the apomict's uniformity arises from the vestiges of its suppressed meiosis. If the chromosomes still occasionally pair they must also cross over; and if they cross over they will exchange different genes which will separate at either of the two meiotic divisions. Since only one division of meiosis can be suppressed in apomixis, there is still segregation at the other division. The progeny consequently shows a muted or muffled variation, a variation so infrequent as to be unpredictable and to justify its having been called 'mutation'.

This type of restricted recombination of differences is probably characteristic of one stage in the evolutionary development of subsexual reproduction, all the stages of which have been found in the grass *Agropyron scabrum*. It accounts for the limited and sporadic instability of many apomictic species in *Hieracium*, *Taraxacum* and *Rubus*, in which every vestige of meiosis has not been lost.

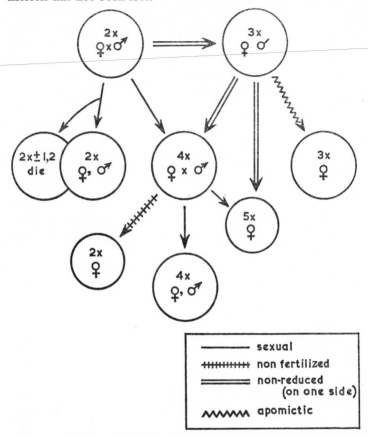

FIG. 21. Versatile reproduction in the dioecious *Casuarina distylis* species complex ($n = 11$) (after Barlow 1959). *Note:* (i) Asexual females apomictic or unfertilized, produce only asexual females; sterile males produce nothing. (ii) $3x$ females by $2x$ males give $4x$ progeny. (iii) The lineage, as in *Poa*, can run down as well as up the polyploid scale.

The second aberration of obligatory apomixis has been described in detail in *Taraxacum*. It is also probably of widespread importance. Here, sexual diploid species have very often given rise to triploids which are apomictic (Fig. 22). These breed true to a great extent; but vestiges of meiosis still remain in the embryo-sac mother-cell. In consequence, chromosomes are lost and 24-chromosome strains give rise in their apomictic seed to plants with 23 chromosomes. These are of eight different types according to which of the 8 chromosomes of the set is lost. The chromosomes may also be doubled to give new 48-chromosome strains. Each new triploid apomictic species thus throws out a little constellation of sub-species. To each of these the systematist may give a name. He will do better, however, to give it, as Sørensen and Gudjonsson have done, a number—a chromosome number.

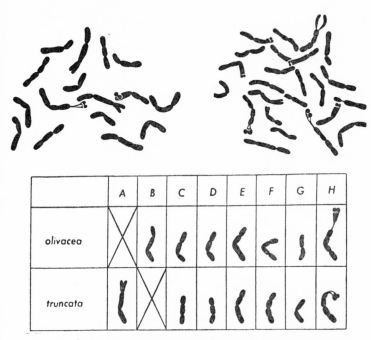

FIG. 22. Chromosome variation in apomictic forms of *Taraxacum*. Above, a diploid sexual form of the *Vulgaria* section and a triploid apomictic form, *T. lacinosifrons*. Below, the extra 7 chromosomes in two of the eight derived ($3x - 1$) types each of which can be named from its leaves or its chromosomes. × 4200 (from Sørensen and Gudjonsson 1946).

In the intervals between mutations in a subsexual strain of plants the strain is constant enough to cause the name-giving naturalist to regard it as a species. In this way many thousands of hawkweed plants of the genus *Hieracium* have been named as botanical species. Ten thousand names attached to minute differences tell us merely that the group is subsexual. The fact that there are diploid sexual species while the apomicts include triploid, tetraploid, pentaploid, and perhaps unbalanced forms as well, tells us the present arrangement or classification and mode of reproduction of the plants to which the name *Hieracium* is given: but it also tell us how the apomicts have arisen and are arising from the sexual species.

6. MUTUAL SELECTION AND DRIFT

Auto-polyploid forms of a species usually show a greater disposition to vegetative propagation, or at least to a perennial habit, than their diploid relatives. This is shown by bulbil-formation in *Ranunculus ficaria*, by root-bud formation in *Biscutella laevigata*, by vivipary in *Poa* species. Similarly, triploid species with no diploid relatives like *Tulipa saxatilis* are remarkable for their habit of multiplying by runners.

Related species in a polyploid series also sometimes have the same relationship. The polyploids may in these cases be auto-polyploids or at least have begun as auto-polyploids. One obvious example is the annual Teosinte *Euchlaena* (or *Zea*) *mexicana* ($2x = 20$) whose auto-tetraploid form is known as *E. perennis*. Another obvious example is the comparison of the diploid sexual sunflower *Helianthus annuus* ($2x = 34$) and the hexaploid vegetatively propagating and perennial Jerusalem artichoke, *H. tuberosus* ($6x = 102$).

The following series are representative of the transition that is found in three genera (from Sakisaka, 1954):

	$2x$ no rhizome	$4x$ little rhizome	$6x$ much rhizome	$8x$ extreme rhizome
CHRYSANTHEMUM	*coronarium*	*indicum*	*morifolium*	*decaisneanum*
POLYGONUM	*orientale*	*japonicum*	*amphibium*	*reynoutria*
DIOSCOREA	*tokoru*	*japonica*	*sativa*	*bulbifera*

So constant is this correlation within species, and so widespread between species, that it might be thought to result from a direct effect of chromosome doubling: but the experimental evidence contradicts such a view.

In experiment, neither auto- nor allo-polyploids tend towards vegetative propagation. We are led, therefore to a radically different conclusion. A constant stream of autopolyploid forms are, as we know, arising in many species. Their sexual infertility filters out for survival those which can reproduce by other means, those which can multiply vegetatively. This they will do when they come from exceptional diploids having the same capacity. Such diploids usually occur even when they have not been found by naturalists, nature having more opportunities than the naturalist for discovering rare and unexpected plants.

The same principle applies to the origin of apomictic polyploids. In *Potentilla*, for example, apomixis is rare in diploids, abundant in the sexually less fertile polyploids. It is facultative when some sexual fertility remains, obligatory when sexual sterility supervenes. Thus sexual fertility, apomixis and vegetative propagation are competing methods of reproduction whose relative advantages are shifted by a change of chromosome number from diploid to tetraploid, which is usually autotetraploid; and even more drastically by a change from diploid to triploid.

This conclusion is systematically most important. For it means, first, that the auto-polyploid forms of a species will always be a sample picked from divergent and unrepresentative diploids, picked by natural selection: and, as we have noted, new tetraploid and hexaploid forms, although cut off from crossing with the diploids, may be continually replenished with new diploid gametes or new tetraploid zygotes produced by the diploids. Secondly, it means that the chromosome constitution and mode of reproduction are *mutually selective*. Any change in the mode of reproduction sets in motion a train of selective effects which in the light of our knowledge of chromosome numbers is no longer concealed from us.

Mutual selection always depends on two genetic variations,

one of which must always have arisen before the other. For example, the capacity for vegetative propagation must be there before the polyploidy. In this respect mutual selection of two genetic properties is analogous to the natural selection of one genetic property in relation to the environment. For it may happen that either the genetic change or the environmental change comes first. If the genetic change comes first, the situation has been described as pre-adaptation.

The principle of mutual selection is the expression at the chromosome level of the elementary notion that two genes A and B, which confer no advantage on an organism separately, confer an enormous advantage when brought together by sexual recombination.

It is this situation which demands the linkage of groups of genes having related effects, thus favouring, as we suppose, their combination by inversions and interchanges which hold together several genes and help to form complexes, super-genes and the differential segments of sex chromosomes.

Most mutual selection is no doubt hard to detect and can therefore lead us into a trap. One of the two components in such a system, it may be polyploidy or interchange, or it may be an invisible gene effect, will often be concealed from the naturalist. He may then conclude that an evolutionary change is non-adaptive and is due to 'drift'. He may conclude that it is due to an entirely non-selective change preserved by an accidental loss of the older alternative genotype in a small population. He will usually be wrong in doing so. Mutual selection is nearly always a component of apparent drift.

Another way of looking at mutual selection appears if we notice that the chromosome constitution and the method of breeding are parts of one system, that is the *genetic system*. It is a system because its parts are adaptively connected: when one part changes other parts are put under pressure to adjust themselves by their own changes.

The different kinds of reproductive mechanism which exist unperceived by the naturalist or systematist, mean that the competition of alternative mechanisms and the action of natural selection are much more oppressive and urgent at the chromosome level than they appear to be at the naked eye level. Our ideal sexual reproduction, the system of what we may call free trade in genes, is always competing with the restrictive practices of vegetative reproduction, true-breeding hybridity, and with other short-term advantages such as those of polyploidy, the exploitation of B chromosomes, and so on. Most sexually reproducing species are seething and surging at the genetic level with these new conditions and practices which are invisible at the external or morphological or phenotypic level. The study of chromosome numbers lifts the edge of the veil which conceals these conflicts. The study of meiosis does much more. And in difficult cases it points the way to the use of breeding experiments which can, of course, reveal almost the whole story.

7. THE ORIGINS OF DISCONTINUITY

Fifty years ago, when chromosome counts were few and largely untrue, it was often supposed that chromosome number must be a character, constant for the members of a species, and distinguishing it from its relatives. This view would imply a virtue and simplicity in species which they unfortunately lack. It would also imply that new species arise suddenly like buds from a parent stock and by a uniform process, that of changing their chromosome numbers. Those who imagined at the outset that such changes would not occur within the species naturally failed to discover them. They thus failed to discover the origins of species.

Darwin, on the other hand, had assumed that all differences which exist between related species arise from, or rather begin as, differences within a species. This assumption is supported by the study (to which we shall return later) of chromosome variation. All the types of chromosome difference which distinguish between species are also found within species. Polyploidy, inversions and interchanges, fragmentation and the possession of B chromosomes, are found alike to distinguish groups within and between species. Most large species include races which differ in one of the first three respects. Not, of course, to the same degree as related species would differ: for all except the B chromosomes these changes hinder the fertile interbreeding of the groups they distinguish. Indeed, chromosome differences which arise within a small interbreeding group of plants usually reduce the fertility of heterozygotes and thus constitute foci of discontinuity, barriers to gene exchange both between chromosomes and between whole populations. In inversion hybrids any crossing-over within the inversion produces fragments whose loss kills the resultant spores or gametes. Therefore, effectively crossing-over is suppressed within inversions. Blocks of genes are fixed.

Thus chromosome changes set events going in new directions. They provide the materials out of which separate mating groups are built and therefore, in the course of time, divergent new forms adapted to different habitats. They establish the genetic isolation which is the complement of Darwin's geographic isolation and mutually reacts with it in the origin of species.

Experience has shown that the point in divergence at which we say one old species has given rise to two or more new ones is a matter of convenience and any rule we make is bound to be arbitrary. Differences of chromosome size, shape and number exist within species and these correspond to the differences found between species and even between larger groups. This is also true, of course, of differences in the external forms of plants. But with the chromosomes the consequences are different because chromosome changes can be mapped; and they can be induced experimentally. Our various means of studying chromosomes enable us to put their changes in an evolutionary sequence which is sometimes certain and otherwise subject to test and enquiry.

No doubt changes which arise in the chromosomes often later become associated by recombination with gene changes and hence with those external differences of form which are recognized by the systematist as the markers for separating groups to which he gives the name of species, genera and so on. They will become associated in this way because the changes in the chromosomes lead to inter-sterility, that is, to genetic isolation and so cause a cleavage in the breeding group of plants.

Comparative studies show in great detail how these events are connected so far as polyploidy is concerned. For polyploidy, although no more important than other chromosome changes (indeed less important in its long term effects), is at once revealed in the chromosome number.

If we examine the chromosomes at mitosis in certain large groups of plants like the species of *Antirrhinum* or *Nemesia*, *Ribes* or *Quercus*, we may find no obvious differences between them in shape, size or number. But more usually, as in *Rosa* or *Triticum*, we find that the different species fall into a poly-ploid series. We also find instances of such series within species, as in *Tulipa*, *Narcissus*, *Solanum* or, above all, in *Galium* or *Mentha*. Here the polyploid forms are so far merely distinct from the rest of the species in some physiological properties which do not appear in the herbarium but express themselves only in their distribution, that is, in their geographical or ecological suitability.

To take an example: there were formerly five species of *Crepis* which are now included as varieties within *C. vesicaria*. But each of these varieties, itself a diploid, has a tetraploid, presumably auto-tetraploid, form or forms in which no varietal distinction has been noticed or allowed. Indeed the type specimen of each may be the diploid or it may be the tetraploid. We do not know although we could tell if the herbarium speci-men had ripe pollen since the tetraploids would have larger grains (Fig. 23).

What happens to new auto-tetraploid forms? Auto-tetra-ploids arising from a diploid species are likely to be continually fed from the parent stock, of whose propagation they are an irregular by-product. They may therefore appear to remain a part of the old species until the two groups move apart geo-graphically. Examples of this divergence are of many kinds. At one extreme is *Prunus cerasus*, the sour cherry, an auto-tetraploid derived from the diploid *P. avium* or sweet cherry stock. The two have moved far apart in external form. But they have remained highly inter-fertile (although their crosses of course being triploid are sterile) and have retained the same centre of diversity in Asia Minor. At the other extreme is the

69

European sea rocket, *Cakile maritima,* whose auto-tetraploid derivative has anticipated or displaced it in America but the two, sharing the same seashore habitat, have diverged little. Indeed, until the distinction by chromosome number showed beyond doubt that the American tetraploid form or species alone exists in Iceland, the European species had always been assumed to have a prior claim on the island. The two forms being similar in appearance but inter-sterile may be described as *cryptic species.*

Sterility barriers establish genetic isolation between tetra-

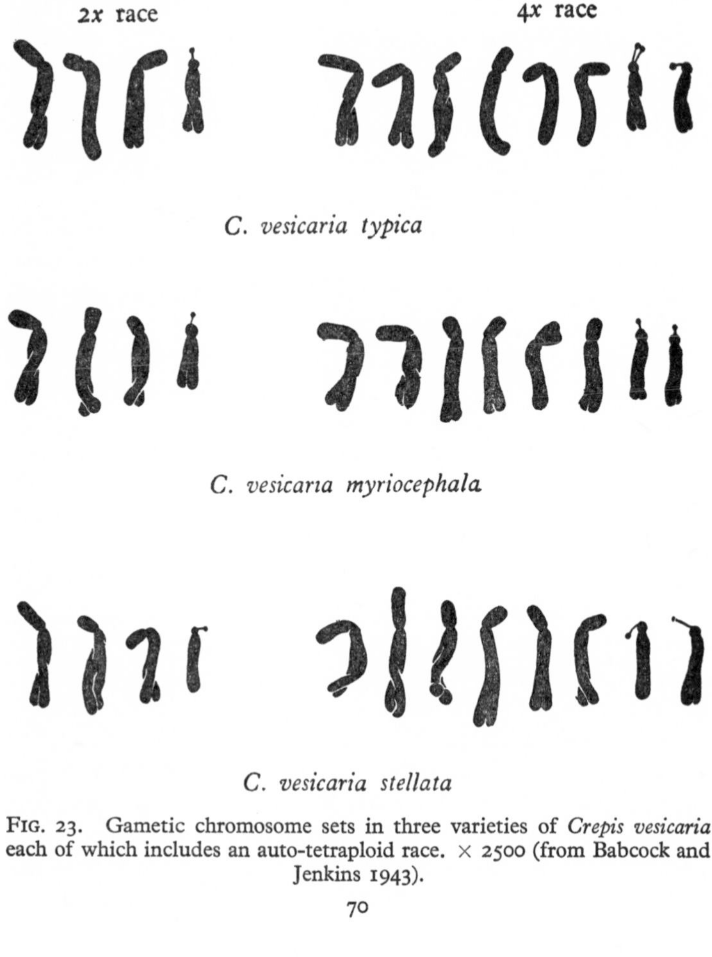

2*x* race   4*x* race

C. vesicaria typica

C. vesicaria myriocephala

C. vesicaria stellata

FIG. 23. Gametic chromosome sets in three varieties of *Crepis vesicaria* each of which includes an auto-tetraploid race. × 2500 (from Babcock and Jenkins 1943).

ploids (or between hexaploids) less easily than between diploids because irregularity of meiosis is less serious. Indeed a hybrid between two auto-tetraploid species will always be, not less fertile, but more fertile than its parents! The laws of species formation for diploids are, as we may say, reversed. We should accordingly expect classification or delimitation to be more difficult in polyploids than in diploids. There are indications that this is so. For example, tetraploids and hexaploids often show ranges of variation in parallel such as would be absurd as between diploids and tetraploids. Thus in the *Achillea millefolium* complex of North America, according to W. E. Lawrence, the following types occur over the range from British Columbia to Mexico:

	dwarf maritime	giant lowland	dwarf alpine
A. lanulosa	$4x$	$4x$	$4x$
A. borealis	$6x$	$6x$	$6x$

Now the types of plant which are capable of crossing are always those with the same chromosome number. But these types differ, and differ in parallel, in respect of habit, form and distribution: the hexaploids merely keep closer to the coast. Thus, in *Achillea* the external and the chromosome differences happen to agree in cutting across the ecological differences together with the genetic differences that these imply.

In the *clusiana* section of *Tulipa*, on the other hand, where differences in flower colour are the ones that attract attention, and profound ecological specialization is not involved, every variation that occurs in diploids occurs also in tetraploids. The external divisions cut across the chromosome ones.

In groups of plants with extensive diversification by polyploidy the systematist relying on morphology alone may thus sometimes reach the right and sometimes the wrong conclusion. He may divide his plants in accordance with their relationship by ancestry or their capacities for progeny; but he may also divide them in conflict with these natural relationships and capacities. His systematic names with their attached authorities and herbarium specimens will then have no connection with chromosome numbers and will fail to tell us what we need to know. Which result he achieves will be a matter of luck rather than of judgment.

The origin of discontinuity is the origin of a species so far as it can be said to have an origin. But since discontinuities are of many kinds, polyploidy merely being the most obvious, so must origins be also. Triploids or tetraploids, although they may arise many times and in many places, must arise suddenly. They are, in their way, final and complete. Discontinuities which spring from inversions, or interchanges, or mere breakage at the centromere, on the other hand, are *cumulative*. One is added to another. They help one another and therefore, over great stretches of time, they are likely to attract one another. In a sense they infect a species, sometimes gradually splitting it into any number of fragments. This is what has happened, as we shall see later, to the species of *Oenothera* as they have moved eastwards across North America.

The same cumulative effects, not in creating, but in breaking down, barriers may well arise from hybridization. The origin of a species (or, as we shall see, of a cultivated plant) is not therefore to be assigned to a particular place and time save in special cases where the genetic and chromosome conditions of the group to which it belongs show that a sudden change has taken place.

8. THE SPECIES IN MOVEMENT

The chromosome study of species, as revealed by our discussion and by a survey of the *Chromosome Atlas*, is only a beginning. But already it radically changes the view of species that is obtained from the use of a systematic flora. Some of the conclusions we reach are obvious as soon as they are pointed out. Others are entirely unexpected.

Examining the chromosomes, and knowing our genetic principles, we are able to recognize species from the single moment, or from the successive moments, of their origin in single cells to the time when they break up into many species or are extinguished altogether. We can see the stages by which they proliferate or disintegrate or degenerate. We can see the grounds for these different changes in the development of their genetic systems, as sexual, subsexual, vegetative or apomictic units.

The results of these different types of organization on the chromosome level and on the external level inevitably interact by mutual selection all the time. Their modes of interaction depend on the structural character at both levels. Chromosomes in different groups have widely different capacities. Their sizes and numbers, their capacities for crossing-over, for misdivision and for polyploidy, their degrees of differentiation, their possession of heterochromatin, and their means of producing B chromosomes, or of generating internal hybridity, all these vary. So likewise do the external generative mechanisms of plants, their seed production, their ability to select pollen and embryo-sacs in fertilization, their capacities for vegetative or apomictic reproduction, for sexual differentiation, for permitting or restricting hybridization.

The most important of interactions are those between different modes of isolation. Spatial and genetic isolation are alternatives in the beginning of species formation. But in all later stages they are likely to assist or reinforce one another; they are likely to be mutually selective. It is worth noting at this point that genetic isolation may take two forms. If it depends on differences in the reproductive organs it may be externally visible and has therefore been used by systematists in classifying and naming species. If it depends on the chromosomes it need not be externally visible; it is not used for naming species and its result is the origin of two or more groups which begin as cryptic species. The situation may be represented as follows:

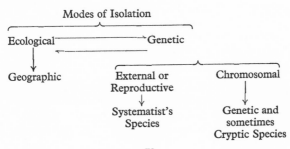

All this means that species are transformed by chromosome studies from the fixed groups of classical systematics to working and moving units, living and changing units. The whole now makes a picture of the kind which the theory of evolution has always required us to imagine but which the naked eye, commonsense classification of living things, has not seemed able to supply.

We now have to see how this dynamic view of the species and of plant-life in general will enable us to understand better the relations of plants with space and time.

III. PLANTS IN SPACE

I. CHROMOSOME ECOLOGY

ALL species with diversified and mutually isolated habitats are more or less adapted to the different conditions under which they live. This adaptation is genetic; it is determined by the constitution of the chromosomes. Moreover, it is of the same kind and has the same genetic basis whether the differently adapted forms are growing near together, when their differences are said to be ecological, or far apart, when their differences are said to be geographic. How do we know this?

When plants of a certain group are taken from different parts of their distribution and we grow them together in good soil, and under favourable conditions in a garden, their range of size between the coastal, alpine and desert types is, as Turesson, Clausen and others have found, largely preserved. It is not entirely preserved, of course, because desert or alpine plants grow better under moderate conditions. The same is true of the variation over their range from temperate to arctic or tropical conditions.

Full and combined genetic and cytological studies have never been made of the variation within diversified species of plants. Studies without much chromosome detail have been made by Turesson on *Hieracium umbellatum* in Sweden, by Clausen and others on *Potentilla glandulosa* in California, and by Gregor on *Plantago major* in Scotland. These workers have concluded, or at least have found no objection to assuming, that in the species they were studying there was a free exchange of genes somewhat as in our ideal diploid species. In this way a variety of genotypes is presented to a variety of environments. Hence groups of plants have been sorted out. These groups differ genetically and are adapted to different habitats. Such groups Turesson describes as *ecotypes*. Where, as in *Plantago*, there is a gradient of habitats there is often a gradient of genotypes, a *cline*. Where, as in *Hieracium*, different habitats and hence different ecotypes overlap, there is hybridization between the ecotypes and by recombination there is a production of misfits. Sometimes, of course, genetic isolation may put a stop to crossing and the separated ecotypic groups will then have the makings of new species.

Now some species, or even groups of species, agree well enough with this picture. In them, it seems, no noticeable changes are taking place in chromosome structure or number. Their differences of genetic adaptation must then be taking place by the mutation and recombination of genes, accompanied perhaps by inversions or translocations of minute segments of chromosomes. Such systems of variation, wholly beneath the visible chromosome level, as it were, may even extend beyond a species. The nearest to establishing this, perhaps, is Peto's study of the cross between the two meadow grasses *Festuca pratensis* and *Lolium perenne*, each with 7 haploid chromosomes. The pairing of the chromosomes is as regular in the F_1 hybrid as in its parents. The frequency of chiasmata is unreduced, but the fertility is low. Only when fertilized with *Lolium perenne* pollen will the hybrid eggs set a few seeds.

The infertility of the *Festuca-Lolium* hybrid is due, not to any failure of the regular mechanism of recombination at meiosis, but to the inviability of most of the regular recombinations of *Lolium* and *Festuca* genes (cf. Fig. 10). If there are differences like inversions distinguishing the *Lolium* and *Festuca* chromosomes—and no doubt there are such differences —they are conveniently disposed in parts of the chromosomes, perhaps near the centromeres, where crossing-over and chiasma formation do not often occur.

The opposite extreme to the invisible basis of variation and of sterility is that in which different ecological and geographical races or species within a group are sharply differentiated in chromosome number. Let us consider a few examples.

Where great variation in chromosome number occurs within a species we have the best possible means of studying the selection of chromosome types by different external conditions. There are two situations in which this variation most easily arises. The first has been studied in the *Erophila verna* complex.

Erophila (or *Draba*) *verna* is a species which contains within it innumerable apparently constant units which were described by Jordan, a French botanist, in the last century as the real species. All these forms or races were sexually reproducing. Indeed they were regularly self-fertilizing. How did they arise from one another? and how did they maintain their independence of one another? The answer to these questions was found by Winge.

In *Erophila verna* different races have a wide range of chromosome numbers. Diploids have evidently given rise to tetraploids and octoploids by mere doubling. These being slightly irregular in pairing have in turn given rise to ranges of new unbalanced numbers. Many of these differing in external form have made successful new types and have found for themselves, as they have spread over Europe, suitable ecological conditions for maintaining themselves: the diploids in very dry places, the near-octoploids in very wet places, and the near-tetraploids in intermediate conditions. Thus in this species changes of balance among whole chromosomes combined with polyploidy have achieved what elsewhere comes about by changes of gene combination. Everywhere the original balanced forms of polyploid seem indeed to have been supplanted; and outside Denmark and Holland the diploid itself (the father of the whole species) has failed to get a foothold (Table 7).

The *Erophila* system is an example of what happens when a species replaces cross by self-fertilization. Each self-fertilizing strain breeds nearly true and thus constitutes a little species, one of the *micro-species* of Jordan. Of course all these types are inter-fertile and when they cross, as occasionally happens in nature, they are bound to throw up a flock of new forms by segregation in the second generation. Variation like that in *Erophila* has also been described in *Cardamine pratensis*, *Saxifraga granulata*, *Cochlearia anglica* and perhaps *Caltha palustris*.

In *Cardamine* (according to Lövkvist, Howard and Banach working on Swedish, English and Polish samples) there is a range of nearly stable or true-breeding types with chromosome numbers from 30 to 84. The lower end of the range again

77

grows in dryer places and is typical of *C. pratensis*, the higher end grows in the wetter places and agrees better with *C. dentata*. Special types of chromosome balance with 44, 50 or 58 chromosomes penetrate into special habitats, for example in the Tatra mountains. Beneath the variation in chromosome numbers is, no doubt, a mutually selected variation in gene contents. Together they enable a complex species to meet a vast range of conditions with continual readjustments. It can recombine genes at the same time that it is unpacking or repacking chromosomes: genes, chromosomes and environment are three independent but mutually selecting variables.

TABLE 7

THE ORIGIN OF SPECIES IN THE *Erophila verna* COMPLEX SHOWN BY THE CHROMOSOME NUMBERS IN 113 SAMPLE PLANTS (after Winge 1940)

2n	2x	4x−	4x+					8x±			
	14	24	30	32	34	36	40	52	54	58	64
Denmark	33	—	22	1	1	11	—	6	—	2	—
Holland	1	—	1	—	—	—	—	—	—	—	1
Britain	—	—	11	—	2	3	1	4[1]	2[1]	—	—
Germany	—	3	—	—	—	—	1	—			
Sweden	—	—	1	—	—	6	—	—	—	—	—
Total	34	3	35	1	3	20	2	10	2	2	1

[1] In Scotland only.

The second instructive situation is that where meiosis and fertilization may be casually and variably omitted. Reproduction is versatile. This, as we saw earlier, is found in several *Poa* species. In each of these species a whole range of forms are found from diploid ($2x = 14$) to decaploid and beyond. In each ovule there are a number of embryo sac mother cells. Some of these undergo meiosis while others form unreduced egg cells. Among these there is free competition between the fertilized and unfertilized development of cells which may or may not have undergone meiosis. The result may be a single seed or twin seeds within the ovule. Moreover, many of the polyploid forms have the additional facility of producing vegetative offsets in place of the florets and thus multiplying viviparously.

These kinds of irregularities are exaggerated in all plants where twin seeds are formed. The second seed is often haploid or triploid. In *Poa pratensis* twins a remarkable result was found by Kiellander. Natural triplets from one ovule of a plant with 72 chromosomes had 42, 40 and 18 chromosomes. Evidently an irregular meiosis had been followed by no fertilization. From the 18-chromosome plant diploids with 14 chromosomes were raised. They passed beyond the limits of *Poa pratensis*: in form as in chromosome number they belonged to *Poa trivialis*. This is thus one of the rare instances of the method of polyploid increase being reversed in evolution. It is also a significant example of one species giving rise to another and fulfilling the fond hopes of our forbears.

TABLE 8

DISTRIBUTION OF CHROMOSOME TYPES OF *Poa alpina vivipara* $(x = 7)$ AT DIFFERENT ALTITUDES IN THE TATRA MOUNTAINS (after Skalinska 1952)

Height $2n = 14$		22	26	28	33–35
500–1100 m	I	17	9	7	2
–1500 m	—	9	9	4	9
–1800 m	—	6	3	5	11
–2300 m	—	I	4	—	9
Total	I	33	30	16	31

Within the range of variation which they normally preserve these species show the kind of adaptation that is found in more stable groups. The lower-numbered forms (which in *Poa alpina* enlarge their variability by the possession of B chromosomes) tend more towards regular sexual reproduction. They also tend towards a more southerly or more lowland distribution. This is well illustrated by observations in Poland (Table 8).

As this species climbs the mountains or moves north it gains chromosomes. The change is not of course due to direct action: it is due to natural selection continually favouring new chromosome complements in newly colonized environments. Nor is it a direct effect of the high polyploidy to increase vivipary and other modes of asexual reproduction. Rather its effect is to reduce sexual fertility and by doing so to give a better chance, a better survival value, to modes of reproduction that cut out the need for a meiosis which is increasingly unsuccessful. In the arctic, *Poa alpina* reaches 57 chromosomes, but above 38 chromosomes sexual reproduction is unsuccessful. Asexual reproduction, therefore, is responsible for extending the climatic and geographical range of the species. But the future of the species, we may be sure, remains, with the sexual core.

2. CHROMOSOME GEOGRAPHY

(i) Polyploidy

In treating of ecological differences, so far we have failed to exclude geographic differences. All geographic differences must indeed begin as genetic differences between plants growing in the same locality. Like everything else connected with the origin of species there is a continuous historical gradation between differences amongst brethren and differences amongst genera. Different requirements of soil, and of climate, and of relationships with other organisms, push the diverging stocks or races, if they are genetically isolated from one another, into different regions. These in turn become the centres of new diversity from which they spread over characteristic distributions in space.

The better defined is this geographical character, the better defined must be the genetic character of the group. A species or a race which is sharply delimited in its geographical distribution is likely to be sharply delimited in its genetic isolation from other races and species. At least so we should expect. But it is the chromosome evidence on which we must rely to demonstrate the principle.

The chromosome situation as it appears before we have

given it its spatial meaning is somewhat as follows. We find in the *Chromosome Atlas* hundreds of examples of polyploidy within species. The general types of arrangements we may put in the following kinds of evolutionary order:

Growth	$2x$	$2x$	$2x$
	$2x, 3x$	$2x, 4x$	$2x, 4x$
	$2x, 3x, 4x$	$2x, 4x, 6x$	$2x, 4x, 8x$
	$2x, 3x, 4x, 5x$	$2x, 4x, 6x, 8x$	$2x, 4x, 8x, 16x$
Decay	$2x \ldots \quad 5x$	$2x \ldots 6x, 8x$	$2x, 4x \ldots 16x$
	$2x \ldots$	$2x \ldots 6x \ldots$	$\ldots 16x$

We find species in nature in all these stages, as well as in every other conceivable combination, of growth and decay. They show all gradations between auto-polyploidy and an allo-polyploidy which is no doubt derived from it by crossing and selection. They also show all degrees of increasing differentiation of form and habitat. Further, we find each of the new forms appearing in all degrees of establishment from that in which a tetraploid (or a hexaploid) is arising independently at many places and failing to perpetuate itself to that in which it is displacing its diploid parents.

The question now arises: What happens geographically in a mixed species? How are the diploid and polyploid forms distributed? The first detailed answer to this question was given by Manton from a study of the European crucifer *Biscutella laevigata*. Here, while diploids are found in the plain, tetraploids have invaded the Alps. They have done so as the ice retreated: and in the highest mountains of Spain (Picos de Europa) even a hexaploid has been found. Can it be, therefore, that new polyploids, auto-polyploids, have a special faculty for colonization, for appearing on the edge of the habitat and for breaking across ecological or geographical barriers which are closed to their diploid parents?

The chromosome geography of *Valeriana officinalis*, which has been examined in detail by Skalinska, enables us to bring two enquiries together and shows what difficulties are removed by doing so. Diploid forms of this species extend across the North European plain into Asia. Into the south of England only tetraploids have penetrated and these have prospered only on dry and limy soils. They have, however, produced a number of octoploid types. The $8x$ forms show a greater diversity than their $4x$ progenitors and the reason is clear. They are incapable of crossing with the $4x$'s. The gene-traffic is always one way and upwards in ploidy. But, through having the same chromosome number, they can cross with quite a different species with which they now come in contact, the $8x$ *V. sambucifolia*. We may represent these comings and goings in the following diagram:

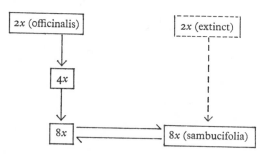

Some of the mixed $8x$ types have invaded wet and acid soils in river valleys. Others have moved on, colonizing a wider area and a wider range of habitats, reaching even to the north of Scotland. Both the $4x$ and the $8x$ forms are, as usual, more prone to propagate by rhizomes and bulbils and are doubtless sexually less fertile than the diploids.

What seems at first glance to be the reverse of these changes has befallen the same complex in Poland. There it has produced, in a parallel way, $4x$ and $8x$ forms. These have moved, not north but south, into the Carpathians: and they occupy, not larger but smaller regions than their progenitor.

Here again, however, it is a case of colonizing new and colder territory, as Europe has grown warmer since the last Ice Age: the new territory happens in this case to have been more restricted than the habitat of the diploid.

A somewhat different situation exists in *Narcissus bulbocodium*. This species has features of its own for it exists in Portugal in the whole gamut of $2x$, $3x$, $4x$ and $6x$ forms, again with B chromosomes in the diploids. They all have a similar size range except that the 6x forms are perhaps the smallest. No doubt the polyploids arise from the dwarfer diploid strains. They probably arise frequently and are of all ages for Fernandes finds some mixed populations of diploids and polyploids. He also finds them close together but in different ecological conditions: and he finds the polyploids occupying alone certain lowland coastal areas, especially with calcareous soils, 'making up the margins of area of distribution of the species'. This is the reverse of the *Biscutella* position in the downhill direction of movement. It is similar merely in the fact of colonization.

On a larger scale is the description by Barber and Stern of the invasion of Europe by tetraploid forms of a whole group of diploid *Paeonia* species, many of which themselves have remained confined to the islands in the Mediterranean (Fig. 24). Here there is a control to nature's experiment in Asia and America, where there is no such sudden barrier to colonization as the Mediterranean Sea. In those continents tetraploids have had no such advantage and they do not occur. As we shall see, in California other means of colonization have been used.

If we admit the principle that chromosome change facilitates colonization we find it leading us to new and unexpected discoveries in plant geography. The evolution of the *eu-farinosae* in *Primula*, as explained by Stern, shows us a gradual increase of polyploidy as these plants have moved both southwards and northwards away from their centre of origin in temperate Eurasia. The diploid species *Primula farinosa*, *P. frondosa* and *P. modesta*, etc., have produced tetraploids (probably auto-tetraploids) which have appeared within their own limits of distribution and variation (e.g. in Gothland). New allo-tetraploid species have colonized the mountains within the diploid range (*P. longiflora* and *P. yuparensis*). Hexaploid and octoploid species have appeared to the north and the south, and the whole group has progressed from 18 to 36, on to 54, 72 and 126 as they have moved into the Arctic and across to the New World (Fig. 25).

Many other groups of plants have passed from the Old World to the New and in doing so have increased their chromosome numbers; or have been able to do so, it appears, only

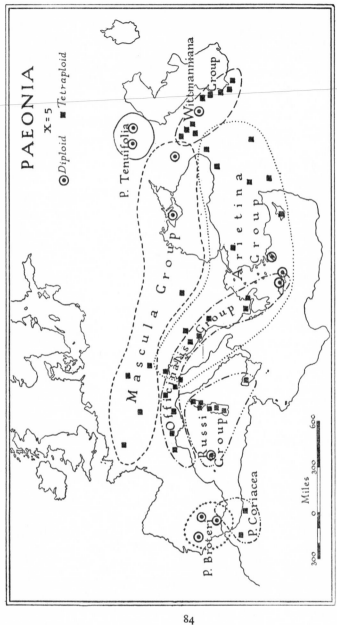

FIG. 24. Map of the distribution of diploid European species of *Paeonia* together with the tetraploids within each of their groups. Note that in the five larger groups the tetraploids are more extended, and in three they are further north, than diploid relatives (from Stern 1946).

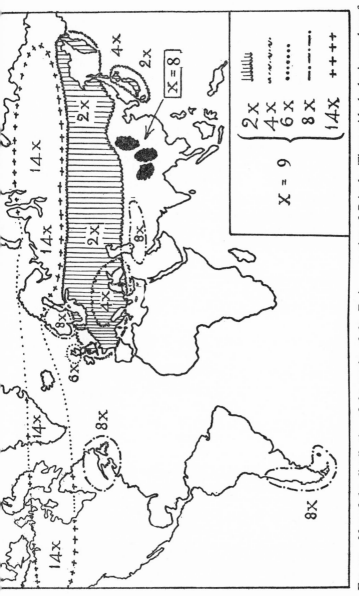

FIG. 25. Map of the distribution of the species of the *Farinosae* section of *Primula*. Those with the basic number of nine show a systematic development of polyploidy with the spread from a temperate Eurasian origin. Those with a basic number of eight are isolated from the rest of the section and show no polyploidy (after Stern 1949).

after such an increase. This is true of *Spiraea* as pointed out by Sax, of *Cakile*, as pointed out by Löve and Löve, of *Iris*, *Gossypium*, *Viola* and many other genera.

A somewhat different pattern is found in *Fragaria* as described by Staudt (1951):

Asia	Europe	N. America	S. America
2x	2x	2x	—
4x	—	—	—
	6x	—	—
		8x	8x

This very simple arrangement is due, it seems, to the polyploidy having arisen in *Fragaria* during the Eocene period separately in the north of the three continents, and the highest polyploids having made their way farthest, even into the southern hemisphere.

Different genera have of course spread under different conditions of time, place and habitat; and their chromosomes have changed in different degrees and in different ways. The best evidence, the clearest record, of how the internal and external conditions are connected, is naturally to be found in genera whose expansion of habitat and diversification of chromosomes have taken place together and have taken place recently.

A special situation arises where extreme or, as we might say, peripheral ecological conditions confront the species in what is geographically the interior of its range. There are many Asiatic genera whose distributions surround the mountainous region of Western China between the headwaters of the Salween and Yangste rivers. This is ecologically the most diversified and florally the richest region in the world. In colonizing new habitats in this territory new polyploid races and species are produced and survive in abundance. Janaki Ammal has mapped the results and shown their consistency in several genera (Figs. 26 and 27).

Thus movement can usually be taken in a merely geographical sense; but there are special regions in which miles are an insufficient measure and ecological colonization is clearly the effective agent in promoting the origin of new species. And if this is true of polyploidy it is no doubt also true of less obvious means by which new species arise.

The kind of region which is being colonized naturally limits the extension of new and changed colonizing types. The polyploid or other diverging type which is climbing a mountain, whether in *Poa* or *Valeriana* or *Rhododendron*, will often have a narrower distribution than its diploid parent in the plain below. But the polyploid which is occupying a large territory made available by the retreat of the ice may well have a much wider distribution; that is until the climatic change is reversed.

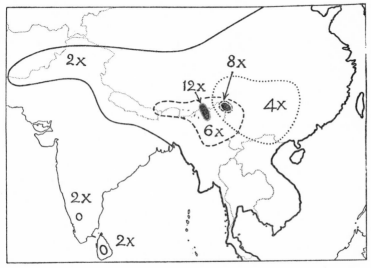

FIG. 26. Map of the distribution of polyploids in Asiatic *Rhododendron* species (after Janaki Ammal 1950).

These examples show how simples rules work when the changes in chromosome number and in distribution are both themselves simple and, as a rule, recent. But when complex movements resulting, for example, from a whole cycle of climatic changes are involved the result is necessarily confused. *Cyclamen* is a good example. Here there is a great range of basic numbers: 10, 11, 12, 15 and 17. The 12 and 15 groups are disposed around Anatolia, the rest around the middle

Mediterranean. In each of these groups we find, at the extreme south, both the lowest and the highest chromosome numbers. At the northern edge, and with the widest distribution, we find the intermediate numbers. This suggests that the diversification of numbers has taken place before as well as after the last Ice Age.

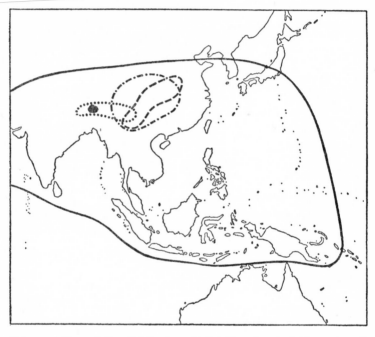

FIG. 27. Map of the distribution of diploid and polyploid species of *Buddleia* in Asia. Whole line, $2x = 38$; broken lines, $4x$ and $6x$; solid, $8x$; dotted, $16x$ (after Janaki Ammal 1954).

Again, as we saw, owing to the more continuous variation in terrain, American distributions are less interesting than European in relation to chromosome numbers. The diagram of Californian distributions (Fig. 28) shows the kind of erratic relations which characterize different genera in the same region. Generally in each form a wide range of habitats is due to a wide variety of races, a wide genetic variation. The higher variation may occur in a diploid or in a polyploid. Or both may be relatively invariable and localized. In this case the higher chromosome number may be in the higher altitudes or the reverse.

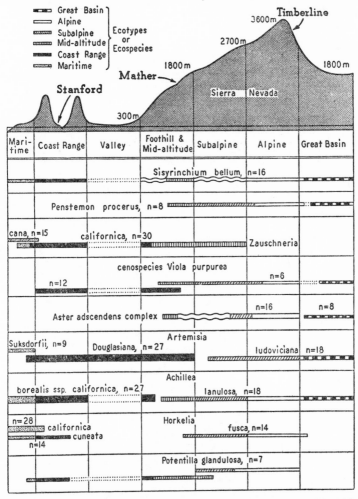

FIG. 28. Patterns of ecological variation within and between species, with and without polyploidy as shown by nine genera in an east-west section through California (from Clausen, Keck and Heisey 1941).

The variety, and indeed the inconsistency of these relationships, depends on gene changes having taken a predominant part in the adaptation of American species. Owing to lack of geographical barriers the species in question have been genetically heterogeneous: they have not therefore been compelled to make use of polyploidy for recent colonization.

(ii) Basic Number

Just as the degree of polyploidy rises when we move away from the earlier home or centre in certain genera, so in others the basic number rises. Here, however, we have the advantage of being able sometimes to survey much greater passages of time than when we have only polyploidy at our disposal. Against this, we may be said to have the disadvantage that the direction of change, by increase, is not so certain or invariable.

The first and the neatest test was found by Stern in *Leucojum*. Here, as the species have advanced from Morocco into northern Europe, their basic number has arisen from 7 to 11 (Fig. 29). For another example, we may take the genus *Crepis* (Table 10). The original basic number is most likely to have been 4 or 5 since the variation has one mode and this is the mode. The place of origin is most probably central Asia. In the genus as a whole we find the highest basic numbers, as well as the most polyploids, in America; and in the section Brachypodes four Asiatic species have $x = 4$, while three Europeans rise to 5 and 6. These facts suggest that the basic number rises in colonizing forms. The same principle also operates within a species. The saxifrage, *Penthorum sedoides*, has 16 chromosomes in the centre of origin of the genus in China. But the same species has 18 chromosomes when it reaches the eastern United States (Baldwin and Speese 1951).

In *Nicotiana* we move on a larger scale both in space and in time. Here the centre is South America where the lowest basic number is 9. In North America numbers up to 12 are found. Only in Australia, with the fewest species and greatest distance, do we find numbers rising to 16, 18, 20, 22 and 24.

In this instance we cannot exclude the possibility of a remote and now concealed polyploidy. Where, however, the change

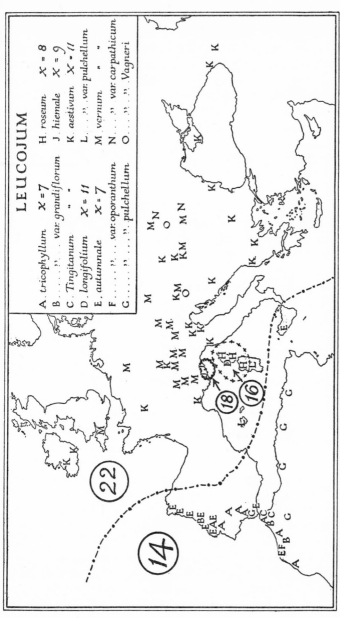

FIG. 29. Map illustrating the colonization of Europe after the ice age by species of *Leucojum* with progressively increased basic numbers (after Stern 1949).

is one of reduction in number, this difficulty can be overcome. Taking the Agavaceae as a whole we have to admit that 30 is probably the original basic number and America the original home. None but the genera most widely departing in basic number from the original one, namely, $x = 19, 20, 21, 24$, have established themselves in the Old World, and only one of these, *Dasylirion*, is found in the New World at all. Thus a change in the basic number, however it may have come about, and whether it involves an increase or a decrease, has been favoured in the colonizing forms.

(iii) Internal Hybridity

There is one special mode of evolutionary change which is not revealed by the chromosome numbers at all but only by behaviour at meiosis: it is the development of interchange hybridity by ring-formation. This method, however, provides the most accurate measure of progressive development that the chromosomes can give; for, as we have seen, it follows an irreversible path which may stretch over a few million years, and most of the steps can be seen in nature. It can also be repeated in experiment, and it can be put on the map to show its geographical meaning.

The *Eu-Oenothera* species are best known. Cleland has found that the chromosome arms are united in 90 of the possible 91 ways in which 14 segments, the two arms of 7 chromosomes, can be combined. Some seven different groups of species have been found in the United States. Those in California are out-crossing and have usually seven pairs of chromosomes. In Colorado, Arizona, and Texas, rings of 4 and 6 frequently appear. To the north and east nearly all species have rings of 14 (or occasionally 12 or 10). These species are almost regularly self-pollinated. But they are heterozygous for their interchanges, and they keep their own character by the elimination of the types with seven pairs which each would produce if these pure types, the homozygotes, did not die (Fig. 30).

On systematic grounds we should suppose that these species had spread eastwards from California. On genetical grounds we should expect them to have changed from crossing to

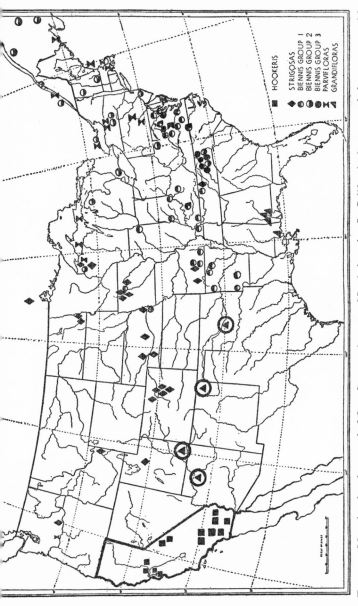

FIG. 30. Map showing the spread of *Oenothera* species across North America. In California there are only homozygous 7(2) species. In the East there are only complex heteroxygote (12) and (14) species. Between, marked by triangles in circles, lie groups of transitional types with intermediate rings (after Cleland 1949).

HOOKERIS

STRIGOSAS
BIENNIS GROUP 1
BIENNIS GROUP 2
BIENNIS GROUP 3
PARVIFLORAS
GRANDIFLORAS

selfing: from the ideal system, with which we expect to begin, to a restricted system. On chromosome grounds we should expect ring-formation to be built up, not broken down. Large rings must be derived from simple pairing, the reverse being difficult or impossible; and, finally, the arrangements of the segments also point this way: Cleland found that the commonest of the 90 arrangements of segments in the eastern species are just those which occur in the homozygotes in California. Thus the unchanged sets of chromosomes have travelled across the continent combining with more and more changed sets.

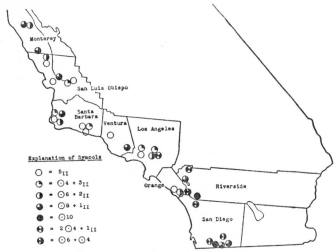

FIG. 31. Map showing the building up of larger rings in *Paeonia californica* as it moves outwards from a centre of origin where the homozygous types with five bivalents are found (from J. L. Walters 1942).

On a smaller scale the analogous situation in *Paeonia californica* points the same way. Here we are dealing with a single species and one confined to California; but within this species, as Walters found, there are the whole succession of developing rings, 4, 6, 8 and 10. We have, therefore, the stages in the unfolding of an *Oenothera* system; and the principle is the same: in the middle lie the homozygotes while to the south and the north more rings and larger rings appear (Fig. 31).

The chromosome change is again associated with movement. Again in *Isotoma petraea* moving westwards across Australia the peripheral colonizing populations are the most altered, the most advanced. But the chromosome alterations, and their genetic implications, are even more fundamental than those concerned in polyploidy or in changes of basic number: for they require the abandonment of sexual recombination among genes and its replacement by a recombination of whole blocks of genes, whole chromosomes, indeed whole groups of chromosomes, as the normal basis of variation within the species.

(iv) The Effects of Migration

The processes and reactions by which complete rings and complete interchange heterozygotes are built up are illuminated by the records of failure. *Clarkia elegans*, related to *Oenothera* and like it beginning in California has the beginnings of a new system as shown in Table 9. But it fails to establish a complete ring and fails to expand. Why?

A number of factors militate against the success of the ring in *Clarkia*. Its annual habit as we have learnt makes the highest demands on fertility. Its chromosome number is a drawback. Probably 18 chromosomes are too many to form a regular ring at prophase or segregate regularly at anaphase. Still more serious is the fact that, from amongst the innumerable new chromosomes arising by interchange and taking part in rings, some give successful trisomics. Not only this, but the extra chromosome, arising as in *Oenothera* by non-disjunction in the ring can be replicated, even up to six times. We thus have a glimpse of the origin of B chromosomes in a species with, we may suppose, an unusually low differentiation of its ordinary set. But such B chromosomes are likely to trip up the ring mechanism. A species can hardly afford to experiment both with large rings and with large B chromosomes at the same time.

The failure to complete the ring leads however to a smaller compensating success. Some of the interchanged rings segregate homozygotes for interchanged complexes. Such segregates arise in experiment with *Oenothera*, especially with extra interchanged chromosomes which remove the lethal effects of homozygotes. But in *Clarkia* they arise in nature. They can be recognized by producing rings when crossed with the original or basic type. These homozygotes appear most often on the ecological margins of the species. Which reminds us that this species, being confined to damp woodland, has ecological margins such as *Oenothera* with its disturbed habitat has never encountered.

TABLE 9

FREQUENCIES OF INTERCHANGE HETEROZYGOTES AND HOMOZYGOTES IN 420 PLANTS FROM 36 NATURAL POPULATIONS IN CALIFORNIA OF *Clarkia elegans* A SPECIES STILL IN A DYNAMIC STATE OF CHROMOSOME EVOLUTION[1] (after Mooring 1961).

	Segmental type in homozygotes[2]	Numbers of Interchanges and Types of Configurations ($2n = 18$)								Total no. of indivs.
		0	1	2		3			4	
		9(2)	(4)	2(4)	(6)	3(4)	(4) + (6)	(8)	2(4) + (6)	
4 Large Popns. 1 2 3 4	A C A A	68 51 14 30	77 31 27 8	9 1 15 —	2 1 2 —	— — 1 —	— — 1 —	— — — —	— — 1 —	156 84 61 38
32 Small Popns. 5–36	28A, 2B, D, E	62	17	1	—	—	—	1	—	81

[1] This species has up to eight replications of normal chromosomes inherited like B chromosomes but not yet correlated with interchange heterozygosity (Mooring 1960)
[2] A is the basic type; in peripheral populations B–E arise as homozygotes from AB, AC, AD, AE heterozygotes.

The development of internal hybridity from interchanges reinforces what we know from polyploids, apomicts, and other evolutionary experiments. These changes meet with every degree of success—short of eternal success. We find them at every stage on their experimental paths. And we have a great variety of evidence from the habit of the plant to the structure of the chromosomes on which we can go in predicting their future progress.

The evidence we have is still fragmentary but it begins to take shape. It seems that the more plants move the more they change. This is true of polyploidy, of internal hybridity, and of basic number. Is it also true of the total gene structure and of its external expression?

There is an obvious reason why forms at the periphery of a group are likely to be the most divergent. In so far as spread is due to colonization of new areas and of less occupied areas where natural selection is relaxed, new and experimental forms will get their best chance when they move, and explore, and invade. Moreover, such selection as does occur is of new and different kinds. This also favours new and experimental forms. Hence change should go with movement where change is possible at all.

The situation is discussed by Ford (in the appendix) in which a species is not expanding but has a stable boundary. Change here also seems to be favoured. It is not, however, by a relaxation but rather by a tightening of natural selection. From this contrast interesting questions arise. Are we concerned in both cases with a reorientation of selection? Is the stable boundary a common alternate phase of the moving boundary? In the most general terms, is it a *selective oscillation* characteristic of boundary conditions which is reflected in the change-with-movement principle?

A third and less obvious consideration is forced on our attention by *Oenothera, Paeonia* and *Erophila*. In them we see the development of special systems of diversification accompanied by the replacement of outbreeding by inbreeding. These changes have taken place on the advancing edge of the species. Why? It is evident that colonizing forms are more likely to be isolated in small populations or as single individuals. Forced to inbreed they are compelled to resort to substitutes for gene recombination such as auto-polyploidy, change of balance, fragmentation, internal hybridity and block recombination. Thus a variety of agencies are used to meet the same kind of genetic crisis.

The principle that it is the most conservative species which will be found farthest from the centre of origin, or centre of diversity, has been maintained for *Crepis* species on morphological grounds by Babcock. This is the very opposite of what our chromosome evidence and our genetic consideration leads us to expect. Can it be that conservatism in external form has nothing directly to do with conservatism in the organization of the chromosomes? An ecological diversity at the centre of a species demands a morphological and a genetic diversity, but it does not so imperatively demand a chromosome diversity. This is a question which the collection of new evidence is likely to answer.

So far as we have gone we have found that the relations of chromosome studies with plant geography, when considered in all their variety, are consistent with the evidence of changing climates and distributions that we obtain by other means. But the details of variation of chromosome number and structure and the consequent adaptive systems differ in different genera. As in treating every other aspect of chromosome systematics, we have to examine the whole situation. We must accordingly now turn from space to time in considering the evolutionary evidence.

IV. PLANTS IN TIME

I. THE FOUR LEVELS OF CHANGE

ALL changes in evolution arise from changes in genes and chromosomes which are followed by differences in the survival and propagation of old and new types. The process of selection responsible for these differences takes place in time as well as in space. Time is often the variable that concerns us. This is so in respect of the direction and the speed of change and also in respect of the kinds of change and the kinds of plant that are changed. In all of these respects the chromosomes give us a new kind of evidence.

Before we examine the evidence we must notice that there are different levels of organization at which we describe evolution. There seem to be four such levels. First, there is the plant itself in its external form, its phenotype. Secondly, there is the sum of the genes which constitute its genotype. Thirdly, there is the arrangement of these genes, their linear totality in the chromosomes; for example it is a similarity of arrangement in mating chromosomes on whose relative permanence in number and constitution the sexual fertility, and hence the continuity, of the sexual species depends: and, fourthly, there is the breeding system, inbreeding or outbreeding, sexual, subsexual or non-sexual, on which the significance of the third level depends. This breeding system, as we have seen, is crucial for every question of systematics and evolution. It is a part of the phenotype and is determined by the genotype. But it is hidden from the naturalist until he has bred the plants experimentally.

Now all these things would be easy to understand if the four levels of evolutionary change were everywhere similarly related. But they are not. The genetic system which they together constitute is itself evolving owing to changes in the relationships of the four levels. This principle governs the appearance of a list of chromosome numbers. It does so in the following way.

99

Change in external form is closely tied to gene change which is its sole ultimate basis. But it is not independent of chromosome change. A single gene may shift a plant, so far as external form is concerned, from one natural order to another. But such a mutated Antirrhinum, for example, cannot reproduce itself and the change is not, therefore, effective in evolution. All profound changes that are to be effective depend on co-ordinated and selected changes of many genes. Polygenes, major genes, and super-genes, or gene complexes such as those associated with interchange in *Oenothera*, are as a rule all concerned. They are shown to be concerned by the study of meiosis and segregation in species crosses. We have no reason to doubt, therefore, that gene change is related, and indeed correlated, with the evolution at the other three levels, which brings into being the species, genera, and families of the systematist. And, as we shall see, the whole of the Angiosperms.

2. BASIC NUMBERS

(i) Simple Polyploidy

The inference of a basic number in a polyploid series is obviously an important step in fitting chromosome numbers to an evolutionary hypothesis. It is by no means always a straight-forward matter. Sometimes the diploid members of the series have disappeared from a genus, or even a family, leaving the evidence of the lower numbers in a remote or unknown place. At the same time the internal evidence of polyploidy may have disappeared also. This is the situation in *Zea*.

The 10 chromosomes of the haploid set in *Zea mays* shows no internal relations, no evidence of an earlier doubling. *Zea mays* is what we call, and rightly call for all practical purposes, a diploid species. But relatives, both in *Coix* and *Sorghum*, have the haploid number of 5. There is a strong suggestion, therefore, that 5 is the ancestral basic number. This number, although of no practical interest for its future hybridization, is the key to the past history, the phylogeny, of *Zea*.

Many even more doubtful estimates of basic number have to be made in surveying the whole body of chromosome numbers. These constitute problems remaining to be solved by more detailed studies of behaviour at meiosis, or of chromosome numbers in missing species, of hybridization or of comparative genetics.

Apart from mere doubling followed by the disappearance of the evidence of polyploidy, basic numbers can change by gain or loss of single chromosomes. If centromeres were of no importance this would be extremely easy. Chromosomes would break or join and the race would be little the worse for it. But basic numbers would be extremely fluid. In the Juncaceae and Cyperaceae an altogether anomalous character of the centromeres puts these principles to the test.

In *Luzula purpurea* the 3 chromosomes of the haploid set can be broken and their parts survive. In consequence the basic numbers of species in *Luzula* (and similarly in *Carex*) form an almost continuous series between 3 and 30 or more, the chromosomes growing smaller as their numbers grow larger. We believe that in these plants every chromosome, instead of having a collective and united compound centromere, has a score of small disunited centromeres. As La Cour has suggested, their organizing abilities are, no doubt, pooled by diffusion of their secretions to the ends of the chromosomes.

This system is restricted to a solitary systematic group of the higher plants, so far as we know. In similarly isolated fashion it also occurs in the Alga, *Spirogyra* and in two animal groups, the bugs and the scorpions. Elsewhere the centromere, specific, localized and with its precise attributes of self-propagation, limits the possibilities of variation in chromosome number. The number can be reduced only by the elimination of a centromere or by the fusion of two terminal centromeres. The number can be increased only by the reduplication, or by the splitting through misdivision, of a centromere. Now terminal centromeres arise, as a rule, only by a misdivision, and this cannot be repeated. We are left, therefore, with little but elimination and reduplication and each of these will involve loss or gain of the piece of chromosome, or perhaps the whole chromosome, attached to the centromere.

These restrictions, as we shall later see in considering

particular shapes of chromosomes, still leave us with room for give and take.

Meanwhile we may note that in about a hundred large genera in the *Chromosome Atlas*, all of them herbaceous, it is possible to make a natural subdivision on a foundation of basic numbers, a subdivision which sometimes, as in *Nicotiana*, agrees but sometimes, as in *Crepis* or *Potamogeton*, disagrees, with the earlier attempts of unguided naturalists (Table 10).

TABLE 10

BASIC CHROMOSOME NUMBERS IN 24 SECTIONS OF *Crepis* (after Babcock and Jenkins 1943)

Section	3	4(8)	5(20)	6	7	11(22)
Zacintha and Phytodesia	3	10	—	—	—	—
7 sections	—	34	—	—	—	—
4 sections	—	13(2)	14(3)	—	—	—
Brachypodes	—	4(1)	1	2	—	—
4 sections	—	—	6	—	—	—
Desiphylion	—	—	1	3	—	—
3 sections	—	—	—	9	—	—
Ixeridopsis	—	—	—	—	3	—
Psilochaenia	—	—	—	—	—	10(A)
113 species	3	61(3)	22(3)	14	3	10

(A) a number of apomictic American polyploids ($x = 4 + 7 = 11$).

Note. Polyploid numbers in brackets. The $x = 8$ species occur along with $x = 4$ species and are therefore assumed to be tetraploid; but the $x = 6$ species do not occur along with the $x = 3$ species and are therefore assumed to be diploid.

Increase in basic number, we might suppose, must always be easier than decrease since a gain of genes does less harm than a loss. Nevertheless, today after more than a thousand million years of chromosome evolution we still find plants, indeed five species of *Crepis*, *Crocus* and *Ornithogalum*, bravely supporting a basic number of three. And the species *Haplopappus gracilis* maintaining itself with only two. How has this come about? There must be long-term advantages in low chromosome numbers which compensate for the short-term difficulties of reducing the chromosome number. The same problem exists with regard to polyploidy. Polyploidy has been occurring since the Pre-Cambrian period. And it is largely irreversible. But about half our flowering plants are still diploids!

The long-term advantages in a low chromosome number no doubt partly depend on efficiency in the management of the key operation of meiosis. Partly, also, they must depend on economy, economy in space in the cell, economy of food in the plant. The chromosomes are the largest structures of the cell and also the most costly piece of machinery that the cell and the organism have to build. In their nucleic acid the plant locks up a large part of the precious phosphorus which it has to absorb from the soil. The fewer genes that can be made to accomplish the organization of development and heredity, therefore, the better the economy, the greater the survival value of the plant. What is remarkable indeed is, not this need for economy, but the outrageous liberties that plants allow themselves to take in this respect, in the interests, that is, of evolutionary experiment.

We must suppose that equilibrium has long been reached in the flowering plants between the selective effects of a short-range ease of increase, and of a long-range advantage of decrease, in the number and substance of the chromosomes. It may be that in the last few million years the important advances have been achieved by reducing the basic chromosome number in Gramineae, Compositae and Papilionaceae. Here the tribes we should put first as having the lowest basic numbers of chromosomes are the ones which the systematist puts last as being the most advanced and specialized in external form.

The rule is, of course, a very rough one since any linear arrangement (however convenient the order may be in treating chromosome numbers or indeed in making any written list) can never correspond with the many-dimensioned processes of evolution resulting from the mutation, recombination and duplication of genes. Nor may we argue the direction of chromosome change in particular instances from the apparent direction of morphological evolution. Primitive forms have not always the earlier numbers. Five is not necessarily, or even probably, the original number of *Crepis* because *C. sibirica* with 5, is the oldest-looking species; and the original Angiosperms had not necessarily, or even probably, 12 chromosomes because that is now the commonest number in Gymnosperms: 7 is much more likely. Chromosomes and organisms are two levels of organization connected by physiology and genetics and only in the simplest cases by arithmetic.

Material properties such as the distribution of hetero-chromatin and the stability of the centromere, or its organization as in *Luzula*, undoubtedly condition the possible changes in basic number. It is not, therefore, surprising that parallel changes in number occur in related genera in the Papilionaceae, in the Gramineae and elsewhere. Such parallelisms may be due sometimes to misclassification of families, genera or species by systematists; and they may be due sometimes to miscounting or mislabelling of species by cytologists. But sometimes parallelisms must be genuine. Their occurrence, like that of morphological parallelisms, warns us of the danger of relying solely on one method in the long range as well as in the short range inference of phylogeny. To the tripod of systematics—form, distribution and breeding—we have added a fourth foot. Later we shall see just where the fourth foot is most helpful.

(ii) Secondary and Dibasic Polyploidy

Change of basic number is something that can be faced more light-heartedly in polyploid than in diploid species. It happens with allo-polyploids in two ways. The first is where whole chromosomes are added to the complement or taken from it to give secondary polyploids with a new balance and a new basic number. Thus in *Dahlia merckii* two pairs of chromosomes have been added to the tetraploid number so that x, which began as 8, has become 18. Similarly, the Pomoideae seem to derive their uniform basic number of 17 from the 7 of *Rosa*.

Thus we may give the secondary gametic number the symbol x_2. If we also give each chromosome a letter we can represent the primary and secondary allo-tetraploids in a diagrammatic way (Table 11).

TABLE 11

ORIGIN AND STRUCTURE OF SECONDARY POLYPLOIDS

DAHLIA: $x = 8$		ROSACEAE: $x = 7$	
Primary D. imperialis* $4x = 32$	Secondary D. merckii $2x_2 = 36$	Primary Rosa $2x = 14$	Secondary Pomoideae $2x_2 = 34$

Primary D. imperialis* ($4x = 32$):

$$
\begin{array}{ll}
A\ A & A_1\ A_1 \\
B\ B & B_1\ B \\
C\ C & C_1\ C_1 \\
D\ D & D_1\ D_1 \\
E\ E & E_1\ E_1 \\
F\ F & F_1\ F_1 \\
G\ G & G_1\ G_1 \\
H\ H & H_1\ H_1
\end{array}
$$

Secondary D. merckii ($2x_2 = 36$):

$$
\begin{array}{lll}
A\ A & A_1\ A_1 & A_2\ A_2 \\
B\ B & B_1\ B_1 & B_2\ B_2 \\
C\ C & C_1\ C_1 & \\
D\ D & D_1\ D_1 & \\
E\ E & E_1\ E_1 & \\
F\ F & F_1\ F_1 & \\
G\ G & G_1\ G_1 & \\
H\ H & H_1\ H_1 &
\end{array}
$$

Primary Rosa ($2x = 14$):

$$
\begin{array}{l}
A\ A \\
B\ B \\
C\ C \\
D\ D \\
E\ E \\
F\ F \\
G\ G
\end{array}
$$

Secondary Pomoideae ($2x_2 = 34$):

$$
\begin{array}{lll}
A\ A & A_1\ A_1 & A_2\ A_2 \\
B\ B & B_1\ B_1 & B_2\ B_2 \\
C\ C & C_1\ C_1 & C_2\ C_2 \\
D\ D & D_1\ D_1 & \\
E\ E & E_1\ E_1 & \\
F\ F & F_1\ F_1 & \\
G\ G & G_1\ G_1 &
\end{array}
$$

* No diploids known.

There are a number of genera in which this kind of secondary polyploid may have arisen by the addition of one chromosome pair to the presumed tetraploid number. The inference is uncertain because one of the evolutionary steps is usually missing. For example, we have in *Saponaria*, (7), 14, 15; in *Amaranthus*, (8), 16, 17; and in *Lavatera*, 7, (21), 22—the numbers in brackets being the missing types.

TABLE 12

THE POLYPLOID DROP

Maydeae: 5–10–9: 18*	*Linum:* (8)–16–15
Cephalaria: 5–(10)–9	*Streptocarpus* ⎫ *Verbascum* ⎬ (8)–16–15
Hesperis: 7–14–13–12	*Vinca:* 8–(24)–23
Bulbine ⎫ *Lepturea* ⎬ 7–(14–13)	*Agavaceae:* 10–(20)–19
Veronica ⎫ *Agrosteae* ⎬ 7–21–20	*Clerodendron:* 12–(24)–23
Camelina: (7)–14–21–20	*Cytisus:* (12)–24–23
Erucastrum: (8)–16–15	*Ardisia* ⎫ *Solanum* ⎬ 12–(24)–23
Cardamine: 8–15–28–30	*Sphaeralcea:* 17–(34)–33

* Tantravahi 1971.

A secondary polyploid of the opposite type arises when the number is doubled or trebled and one chromosome is dropped. The original diploid, or its straight tetraploid or hexaploid derivative, has often, perhaps usually, disappeared. This change must, as we saw, be due to loss, whether directly of a whole chromosome or as in *Cardamine pratensis*, by fusion of

TABLE 13

A. Herbaceous Plants

(i) *BRASSICA* ($x_2 = 9 + 10 = 19$) Known in Cultivation
$\quad$ *B. oleracea* $\times$ *B. napus* = Swede Turnip

$$2x = 18 \qquad 2x = 20 \quad 4x = 2x_2 = 38$$

(ii) *NARCISSUS* ($x_3 = 7 + 7 + 11 = 25$) Known Species
$\quad$ *N. juncifolius* $\times$ *N. tazetta* = *N. dubius*

$$2x = 14 \qquad 2x = 22 \quad 6x = 2x_2 = 50$$
(unreduced egg)

(iii) *CREPIS* ($x_2 = 4 + 7 = 11$) Known Groups of Species
$\quad$ *C. pulchra* group : *C. nana* group : *C. occidentalis* group

$$2x = 8 \qquad\qquad 2x = 14 \qquad 4x = 2x_2 = 22$$

(iv) *VIOLA* ($x_3 = 6 + 10 + 11 = 27$) Known Subgenera
$\quad$ *V. pubescens*, etc. : *V. cornuta*, etc. : *V. arvensis*, etc.
$$2x = 12 \qquad\quad 2x = 22 \qquad 4x = 2x_2 = 34$$
$\quad$ *V. arvensis*, etc. : *V. odorata*, etc. : *V. affinis*, etc.
$$4x = 34 \qquad\quad 2x = 20 \qquad 6x = 2x_3 = 54$$

(v) *VERONICA* ($x = 7, 8, 9; x_2 = 15, 17; x_3 = 26 = 8 + 9 + 9$)
$\quad$ Known Subgenera

B. Woody Plants

(vi) *LOGANIACEAE*————Known Genera (see fig. 34)

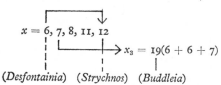

$$x = 6, 7, 8, 11, 12$$
$$\longrightarrow x_3 = 19(6 + 6 + 7)$$

(*Desfontainia*)$\quad$(*Strychnos*)$\quad$(*Buddleia*)

(vii) *MAGNOLIALES, LAURALES:* Known Orders

	$x = 7$(all $4x$)	$x = 12$	$x_2 = 19(7 + 12)$
LAURACEAE	—	*Cinnamomum*, etc.	—
SCHIZANDRACEAE	*Kadsura*, etc.	—	—
WINTERACEAE	*Illicium*, etc.	—	—
TROCHODENDRACEAE	—	—	*Trochodendron*
MAGNOLIACEAE	—	—	*Magnolia*, etc.
CERCIDIPHYLLACEAE	—	—	*Cercidiphyllum*, etc.

two with the loss of a part of one containing a centromere (cf. Fig. 36). Many examples of gametic numbers (those in brackets being missing steps which are merely inferred) show what appears to be such a drop (Table 12).

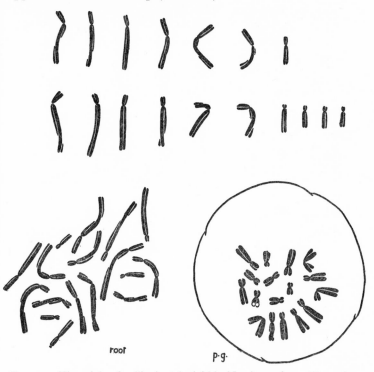

root

p·g·

FIG. 32. The origin of a dibasic polyploid in *Narcissus*: above, *N. poeticus*, $n = 7$, *N. tazetta*, $n = 10$; below, root-tip and pollen grain in the hybrid 'Geranium', $2n = 17$. *Note:* the unreduced pollen grain is the only kind that lives, hence new races and species can arise with 34 chromosomes. × 1200 (from Wylie 1952).

This characteristic change is no doubt due to the fact that the polyploid is able to afford a loss that would ruin a diploid. And the loss proves to be an evolutionary advantage.

An even more striking chromosome number relationship is that where two different basic numbers are combined in one polyploid. The two numbers are added together to give a new dibasic one which we may again call x_2 (or tribasic and x_3 if three sets are added together). This situation is found in all evolutionary stages from that in which it can be repeated experimentally, the old and new types being within the limits of one generous species, to that in which it is already embedded in the relationships of large and ancient groups. We may indeed arrange our examples in an evolutionary sequence, as in Table 13.

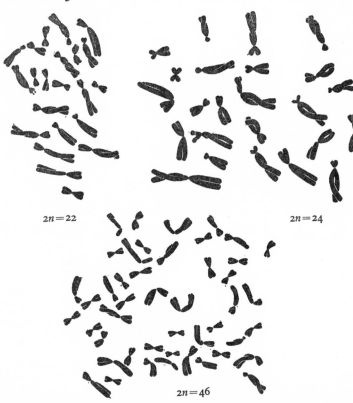

FIG. 33. The relations between a dibasic polyploid and its diploid relatives in the Amaryllidaceae. Above, *Hippeastrum solandriflorum*, $2x = 22$; and *Leptochiton quitoensis*, $2x = 24$. Below, *Hymenocallis littoralis*, $2x_2 = 46$. Note the reduction to half size of the chromosomes of the polyploid species (an adaptation also known in polyploid species of animals). × 1300 (Snoad unpub.).

In the newer cases of *Brassica* and *Narcissus* (Fig. 32) we can point to the precise species which seem to have begotten the new hybrid and dibasic polyploid. But in *Crepis*, *Viola*, *Veronica* and *Hymenocallis* (Fig. 33) this is not of course possible, and in the woody *Loganiaceae* and *Magnoliales* even the ancestral genera have vanished: we can point only to their number-bearers.

In *Crepis* the inference is as follows. We know that 4- and 7-chromosome species occur near the 11-chromosome species. We know that allo-polyploid species like *C. crocea* (Fig. 11) can arise in *Crepis*. And, finally, we know that the chromosome numbers in the genus show a gap between 7 and 11 (Table 10) which can be bridged by the effect of crossing 7 with the commonest number, 4. We may thus visualize such a sequence in changes of basic number in *Crepis* as the following:

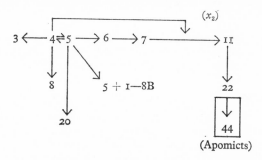

Turning to the larger groups and the remoter inferences, in the Magnoliales and Laurales where the basic numbers do not seem to have changed since the Cretaceous period, we may suppose that the groups with 19 chromosomes are derived from crossing species with basic numbers of 7 and 12 chromosomes which are still found in these orders. It is a hypothesis that can be tested by further chromosome counts.

The dibasic polyploid has bearings beyond the convenience of dividing genera or tracing the phylogenies of species and families. Like other inferences from the chromosomes it bears on the principles of the subject; for every dibasic polyploid must be derived from the crossing or hybridization of two diploid races, or species, or genera, which differed in basic number. Where these ancestral types still survive there can be no doubt that the origin is polyphyletic. And if it can be polyphyletic where the basic numbers of the ancestors differed, it can also be polyphyletic where they were the same.

The conclusion derived from experiment that the polyploid species with the best expectation of life should be derived from crossing widely different diploid species is thus strengthened. The pattern of descent is not merely divaricate: it is also reticulate. And the amplitude of the network can be shown by the chromosome complement. It can be shown in no other way; and it can be shown sometimes after a few years, sometimes after millions of generations.

3. CHROMOSOME INDIVIDUALITY

(i) Variation in Size

If the size of chromosomes changes in the course of descent it may obviously be the result of fragmentation or fusion. This is clear from close experiment as well as from remote comparison. But it can also happen in another way. After special treatments (such as X-raying), as well as spontaneously, we find whole mitoses or even whole buds with smaller chromosomes. Not one chromosome but the whole complement is uniformly smaller. This is clearly not due to a structural change, to the breakage and reunion of the strings of genes, but to a change in the control of mitosis as a whole. The point in propagation at which the polynemic bundles of DNA become unstable and split into two must have occurred prematurely to make the chromosomes enter mitosis at half-size.

When a change in chromosome size occurs after treatment no doubt the shock of treatment has been responsible and the change may be reversible. Such changes in polynemy must occur often enough to make possible a fairly ready adjustment of chromosome size to the needs of a species.

This adjustment occurs sometimes for reasons we do not yet understand. We do not yet know, for example, why *Euphorbia monteiri* should have chromosomes a hundred times

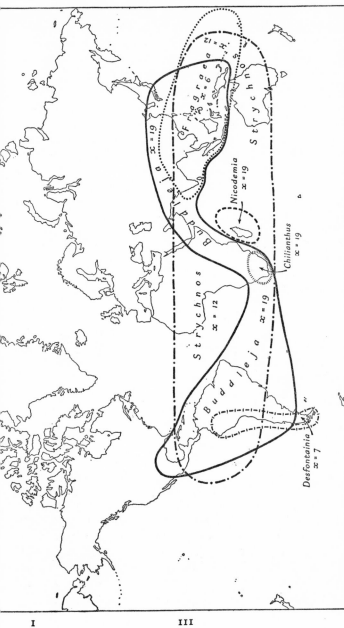

FIG. 34. Geographical and chromosomal evolution in the Loganiaceae. Tetraploids based on six have combined with diploids based on seven (now nearly extinct) to give a dibasic group of three genera based on nineteen (from Janaki Ammal 1954).

larger than *E. micrantha*, the next species in the list. Or why *Trillium* should have chromosomes a hundred times larger than its supposedly close relative *Medeola*; or *Drosophyllum* a thousand times larger than *Drosera*. In certain genera, however, the causes at work can be ascertained.

Within the genus *Crepis* the size range is of the order of ten-to-one. Stebbins and others have noted that the smaller chromosomes are found in the species of *Crepis* with smaller plants and with an annual habit, especially those living under extreme conditions (Fig. 35). The same is true, according to Rees, in *Lathyrus*. The smaller chromosome complements are then seen to be reduced replicas of the complements of their normal brethren (Fig. 35).

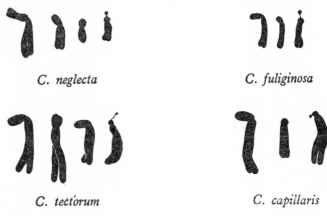

C. *neglecta*

C. *fuliginosa*

C. *tectorum*

C. *capillaris*

FIG. 35. Large and small chromosome species in the 3-group and the 4-group of *Crepis*. The size ratio is probably 4 : 1. × 2500 (from Babcock and Jenkins 1943).

Even more instructive is the evidence from comparison within the polyploid series of one genus. In *Dianthus* and in *Chrysanthemum* it is regularly found that the polyploid species have smaller chromosomes than the diploids. In some species the reduction in size of the chromosomes, like that of the whole plant, has certainly preceded the polyploidy. It has been a case of pre-adaptation and mutual selection. For example, the only natural hexaploid in *Narcissus* occurs in *N. bulbocodium*, the only pentaploid in *Tulipa* in *T. clusiana*. In both genera these are the species which are pre-adapted by having the smallest chromosomes; and no doubt the diploid individuals in these species which gave rise to the polyploids were those which were pre-adapted by being plants of the smallest size. Again, tetraploid species occur in several sections of *Tulipa*, but in *Lilium* and *Tritillaria* where the chromosomes are the largest among plants, triploidy is the limit. Tetraploidy in *Lilium*, like hexaploidy in *Narcissus pseudo-narcissus*, arises and survives only under the protected conditions of experiment and cultivation.

Chromosome size can be shifted up or down, doubled or halved, during development both in normal tissues and abnormally after treatment of pollen grains. It is merely a question of breaking the connection between replication and splitting of the chromosomes in the resting nucleus. Thus it seems that evolutionary reduction in chromosome size is an adaptation, on the one hand to a decrease in the size of the cell or, on the other hand, to an increase in the number of chromosomes.

These observations are consistent with the chromosomes being based on multiple threads, their degree of *polynemy* and hence their size being an adaptive character of the whole complement.

The smallest chromosomes are found in hard-wooded plants whose small cambial cells presumably restrict the room for mitosis. The largest chromosomes appear only in a few groups of the flowering plants: in the Gymnosperms, in the Ranunculaceae and Berberidaceae, in the Gramineae, in the Alismataceae, and in the Liliales. But in each of these groups we find small-chromosome types as well, and if carefully surveyed they might tell us much of the connections of chromosome size with habit and habitat, structure and classification.

What these connections may be we learn from a study of chromosome size as measured by the DNA content in 300 species of Angiosperms. The small chromosome species have the more rapid mitoses and more rapid maturity. They alone are capable of a short life cycle. The large chromosome species are predominantly those which we classify as obligatory perennials. The rate of production of DNA necessary for the reproduction of the chromosomes is thus successively a limiting factor in the growth of the plant and in the evolution of the species.

(ii) Variation in Shape: Bimodality

In most groups of plants and animals there is either a moderate uniformity in the sizes and shapes of the chromosomes or an even gradation of types within the complement. The comparison of chromosome types of different species in these circumstances is profitable only when, as in *Crocus* or *Crepis*, the chromosomes are fairly large, and even then it does not reveal the exact processes of change. Sometimes it happens, however, that within one set there are two types of chromosomes which are sharply contrasted in size and shape; or, as we may say, frequency distribution in chromosome size is bimodal. Bimodality is of course characteristic of the complement which includes B chromosomes. But here we are speaking only of the honest indispensable set of A's. The most instructive instances of true bimodality arise in what were the old groups of Liliaceae and the Amaryllidaceae.

One method of change may be recognized where we have two types of chromosomes which we may call V's and I's: the V's have two large and equal arms with a nearly median centromere, the I's have a nearly terminal centromere so that there is only one large arm about the same length as that of a V. In the Asiatic genus *Lycoris* we find the following types:

L. sanguinea	$n = 11$:	II	II	II	II	II	I
L. straminea	$n = 8$:	V	V	V	II	II	I
L. aurea (Formosa)	$n = 7$:	V	V	V	V	II	I
L. aurea (Japan)	$n = 6$:	V	V	V	V	V	I

In addition there are hybrid types such as would arise by crossing, and there are triploid vegetative species. Clearly this series represents a uniform type of change in which pairs of I's have, in effect, fused to give V's or V's have broken in half to give pairs of I's.

The same relationship exists in *Lilium* and *Fritillaria*. All the species of *Lilium* have 2 V's + 10 I's ($n = 12$). This is also true of most species of *Fritillaria*. But *F. pudica* has 1 V + 12

I's (i.e. $n = 13$) and *F. ruthenica* and *F. nigra* have 5 V's + 4
I's (i.e. $n = 9$). Here we can scarcely doubt that $n = 12$ is the
original type and that both fusion and fission have taken place
to give the diverging numbers (Fig. 36).

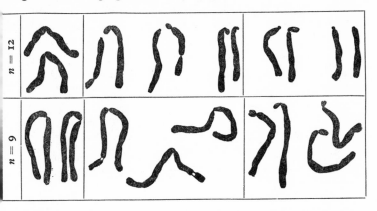

FIG. 36. The haploid sets of species of *Fritillaria* to show evolution by
fusion of chromosomes. Above, *F. latifolia* with 2 V + 10 I; below, F.
ruthenica with 5 V + 4 I. Note the nucleolar constrictions are in two I chromosomes which enter into two V's. × 1400 (after Darlington 1929).

How have these changes taken place? Successive interchanges of segments of chromosome near the centromere with
losses of small fragments would account for the apparent fusion.
The fissions, on the other hand, might occur more directly by
misdivision at the centromere itself giving rise to telocentrics
(as seen experimentally in *Campanula*) and hence to iso-
chromosomes which may lose most of one arm: thus a V could
become replaced by 2 I's. This seems to have been the method
in *Alisma* and *Tradescantia* (Figs. 37, 38).

Whatever the process, its results, we must suppose, are
vigorously selected. Many unsuitable types of chromosomes
must be formed. But they are eliminated. In each group of
plants cell mechanics or gene physiology allow only a limited
pattern of shapes. This restriction is most brilliantly demonstrated in another group or system. This concerns the desert-
living genera formerly in the Liliaceae and Amaryllidaceae

which Hutchinson has joined together in the Agavaceae. It also concerns their relations with the Aloinae and a number of other groups which still remain in the Liliaceae.

The chromosome complement in these plants, as McKelvey and Sax first pointed out, has a unique character. The chromosomes are of long and short types, L and S. The L chromosomes

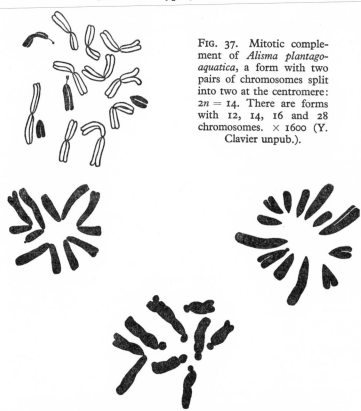

FIG. 37. Mitotic complement of *Alisma plantago-aquatica*, a form with two pairs of chromosomes split into two at the centromere: $2n = 14$. There are forms with 12, 14, 16 and 28 chromosomes. × 1600 (Y. Clavier unpub.).

FIG. 38. Changes in the chromosome complement due to fusion or splitting near the centromere in the *Tradescantiae*: metaphases of mitosis in the pollen grain. *T. gigantea*, $n = 6$, × 1600; *Tripogandra disgregea*, $n = 8$, × 2800 (after Darlington 1929); *T. micrantha*, $n = 13$, × 1500, nearly all telocentrics (after Anderson and Sax 1936). *Note:* (i) each complement has a nearly uniform shape; (ii) the chromosomes of *T. disgregea* have been reduced to about 1/8th of the standard size in this group of species.

are J-shaped such that their long arms are more than five times as long as their short arms or as either arm of an S. The S chromosomes look as though they had come by misdivision from the short arms of the L chromosomes (Fig. 39). There is a complete absence of the V chromosomes which make up the entire complement in many plants such as *Tradescantia gigantea* and account for one or two members of the complements of nearly all others such as *Hyacinthus orientalis* or the *Lilium* and *Lycoris* we have just been discussing. The situation is summarized in Table 14.

TABLE 14

BIMODAL CHROMOSOME COMPLEMENTS IN THE *Liliales: Agavaceae*
AND *Liliaceae* OF HUTCHINSON

AGAVACEAE

(i) *Amaryllidaceae* of Engler

Agave
Beschorneria } $x = 30$:—5L + 25S
Fourcroya

(ii) *Liliaceae* of Engler
Yucca $x = 30$:—5L + 25S
Dracaena $x = 19$
Phormium $x = 16$ } size difference reduced
Doryanthes $x = 12$

LILIACEAE

Hosta (or *Funkia*) $x = 30$:—5L + 25S
                              (2S's are intermediate)
Eucomis $x = 15$:—4L + 11S
                (one L with long short-arm)
Aloë
Gasteria } $x = 7$:—4L + 3S
Haworthia
Hyacinthus amethystinus $x = 14$:—4L + 10S
Ornithogalum } isolated species have a reduced size difference
Scilla

Let us look at the question from the point of view of one wishing to ascertain relationships by descent. The alternative possibilities are (i) that the bimodal or *Agave* type of complement has arisen separately, that is in parallel, in these different groups of plants and (ii) that they have a single origin

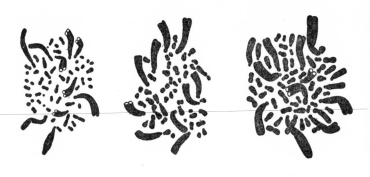

Yucca   *Agave*   *Hosta*

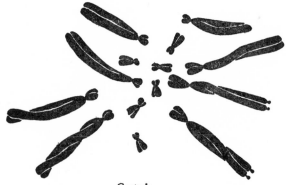

Gasteria

Eucomis   Fig. 39.   *Ornithogalum*

118

and constitute a group of common descent separate from all others. The answer is given by the joint consideration of the morphology, physiology and chromosome numbers of the groups, as follows:

  (i) It is at once clear that Hutchinson's union of the *Yucca* and *Agave* groups is vindicated.

  (ii) It is also probable that the *Aloë* similarity of habitat and chromosome complement and the *Hosta* identity of complement arise from common ancestry with *Yucca* and *Agave*.

  (iii) It is possible that the same applies more remotely to *Eucomis*, an ancient polyploidy having intervened between its type of complement and that of *Agave*

  (iv) Finally, it seems that an *Agave*-like complement may be arising independently in parallel in such genera as *Hyacinthus*, *Scilla* and *Ornithogalum*, a parallelism arising merely from a remote genetic relationship.

How are the *Aloë-Eucomis-Agave* series of numbers related? The easiest explanation is perhaps a remote polyploid increase followed by a paring away by loss of fragments from some of the L chromosomes in the polyploids. Within the *Yucca* group we find another kind of change. Where the basic number is reduced there is a reduction of the size difference and the conversion of I types into V types as though S chromosomes had fused at the centromeres. With this direction of evolutionary change by reduction of basic number, it would then follow, as we saw, that the genera *Phormium* in New Zealand and *Doryanthes* in Australia, which have travelled farthest

---

FIG. 39. Types of bimodal chromosome complements from root-tip mitoses in the Liliales:

Yucca aloifolia ⎫	
Agave americana ⎬	$2n = 60$
Hosta lancifolia ⎭	
Gasteria bayfieldii	$2n = 14$
Ornithogalum umbellatum	$2n = 54$
Eucomis bicolor	$2n = 30$

Note that the sizes of the chromosomes are comparable except in *Yucca*. All from Matsuura and Suto (1935) except *Gasteria* which is by Snoad (unpub.). × 2400.

geographically from the centre of the group in Mexico, show most change in chromosomes. Migration and change are again correlated.

The series of evolutionary changes suggested here is therefore:

$$7 \text{———(14)———} 15 \text{———} 30 \text{———} 19, 16, 12$$
$$\text{(doubling)} \qquad\qquad \text{(doubling) (loss)}$$

Superimposed on these long-term changes is, of course, the usual short-term intra-generic and intra-specific polyploidy.

This characteristic pattern with bimodality for shape as well as for size, like the basic number of the Magnoliales or the Pomoideae, has evidently remained stable over several geological periods of evolution. But strangely enough a similar pattern is characteristic of a somewhat larger and much older group of animals. Certain reptiles and all birds form a group distinguished in this respect from other reptiles and all mammals. They must have remained stable in regard to bimodality of chromosome shape since the Palaeozoic period. And the mammals and birds must have been derived from groups of reptiles distinct in this respect since that time.

In these instances the pattern of the complement, including the chromosome number, is in a sense a genetic character, a character of immense weight and one that is constant, not for a species or a genus, but for a whole family or class and for a large fraction of geological and evolutionary time.

4. EVOLUTIONARY PRINCIPLES

(i) Types of Species

In all organisms, so experimental evidence shows, the chromosomes are similarly liable to undergo inversions and interchanges. They may even, although this is less general since it depends on the centromere, be equally liable to change in number. But the effects of this liability, the consequent changes which are capable of survival, are widely different in different groups.

Festuca pratensis and *Lolium perenne*, as we saw, show no evidence of changes in chromosome number or shape distinguishing the two genera. Their hybrids at meiosis, moreover, show no evidence of interchanges and inversion crossing-over must be infrequent. On the other hand, hybrids of different *Lilium* species show evidence of many inversions; and hybrids of races of *Datura stramonium*, and of species of *Oenothera*, show evidence of many interchanges. In all these groups the structure of the chromosomes changes but their numbers are constant. In yet other genera a variety of chromosome numbers are found. Sometimes, as we have seen, this is due to polyploidy and sometimes to change of basic number and sometimes to both.

Again, within the species, differences of basic number are found revealing cryptic species where the external form shows no differences at all. Examples are *Alisma plantago-aquatica*, *Trifolium subterraneum* and *Vicia sativa*. In many other species, as we have seen, a dozen or more chromosome numbers can be found within the limits of one species and without any change of basic number. These differences may arise in many different ways, from polyploidy, from unbalanced numbers or from the possession of B chromosomes, and sometimes, as in *Narcissus bulbocodium*, in all these ways together.

These differences in effective stability of chromosomes and of complements are important both for their cause and their consequences. Their cause is, clearly, that from an enormous and varied supply of kinds of change each group of plants selects[1] a highly specialized sample, rejecting the greater part. And sometimes it rejects all visible chromosome changes whatever and depends entirely on gene changes which are invisible in the cell.

The consequence of these differences is that chromosome variation can give us entirely different evidence of evolutionary processes in different groups. Where the chromosomes are less stable they tell us what is happening at this very moment to races and species. Where they are more stable they tell us what happened in former geological and evolutionary epochs. In

[1] Or, more strictly in Darwin's phrase, 'nature' selects, but organisms may here be said to act on behalf of nature, possessing a power of attorney in respect of their own chromosomes.

the first situation they tell us how one old species may split up into several new ones and how the new ones are related in ecology or in geography. But in the second situation they label whole tribes, families or orders, those large groups in fact whose relationship by descent since the Cretaceous or even the Permian period puts to the systematist questions some of which can now evidently be answered by chromosome study and in no other way. The chromosomes provide us with a record of the past, a living record, significant in a surprisingly similar way to the dead record which fossils provide for the palæontologist.

(ii) Adaptive Rates of Change

If the greater part of variation, and even most kinds of variation, are rejected in each group of plants we have to avoid certain very obvious assumptions about the relationships of variation at different levels. We must ask again two main questions. First, does stability in external appearance mean stability in genes and chromosomes? Secondly, does progress, that is change from a primitive to an advanced level, in external form imply a parallel progress in the chromosomes?

The answer is that both these assumptions may sometimes be right but that neither can be rigorously justified. The evidence is scanty. One of the few indications is the statement of Janaki Ammal that the fossil *Magnolia* species of the Cretaceous and Eocene correspond with diploid species now living, and not with polyploids. The lines or stocks which have changed most in chromosome number, in this instance, have also changed most in external form. On the other hand, we need not suppose that an animal whose external form, as shown by fossil remains, has been unchanged since the Cambrian period has kept its chromosomes and its genes unchanged in the interim. The two forms would surely no longer be inter-fertile.

The gene and chromosome mechanism, the whole genetic system, is itself adapted to securing change and adaptation of the external form. But if the external form does not need change or adaptation, the genetic system cannot be safely or easily frozen. The pressure of variation will remain but the great power of selection will be directed, not to producing change, but to maintaining the *status quo*. Our problem is therefore the reverse of that faced by the nineteenth century. Our greater difficulty is to account not for evolution, but for stability. Our problem is not to explain the occasional origin of new species but rather to account for the occasional constancy of old ones.

TABLE 15

REVERSIBILITY OF EVOLUTIONARY CHANGES[1]

Basis	Change
GENES	Polygene $\rightarrow$ Major Gene $\rightarrow$ Super-Gene Primary $\longrightarrow$ Secondary $\longrightarrow$ Gene Complex Inversion Inversion
CHROMOSOMES polyploidy	$2x \longrightarrow 4x \nearrow^{3x} \searrow \longrightarrow 8x$ $\longrightarrow$ Secondary Polyploids
basic number	$x = 7 \underset{\longleftarrow}{\longrightarrow} x = 8, 9, 10$
BREEDING SYSTEMS (Gene + Chromosome)	Sexual $\longrightarrow$ Asexual or Apomictic $(2x, 3x, 4x)$ $\rightarrow$ Subsexual $\longrightarrow$ $\longrightarrow$ Complex Heterozygote (ring-forming) Hermaphrodite $\rightleftarrows$ Dioecious Monoecious $\rightleftarrows$ Dioecious

[1]See also Section (iv), p. 127.

External stability (and *a fortiori* irreversibility in evolution) no doubt arises from the interlocking of the three levels of evolutionary change seen in Table 15. It is easy enough, as we saw, to get a new individual that will work as an individual. It is much more difficult to get it to breed. For this the three levels and all the interlocking stages of the sexual cycle have to work together and in the right sequence. How difficult this is we see from the fact that every cell with an abnormal chromosome number is eliminated in the development of every plant. This is a process of natural selection which reaches its climax following irradiation. For every cell may then be abnormal; every cell is then eliminated and the plant or animal dies.

The plant, or rather the race of plants, has to keep a continuity and mutual fitness between one generation and the next if it is to guarantee its sexual fertility and evolutionary survival. This principle (which accounts for a large part of the wastage of variability in nature) should apply, without a strict correlation, both to the genotype (as controlling external form) and to its visible vehicle, the chromosome. But where sexual generations are lengthened we might expect the restriction to act more severely on the chromosome mechanism. Fertility is more easily upset by chromosome changes than by changes in genes one at a time.

We might expect that in long-lived plants chromosome numbers would change less than in short-lived plants which were evolving and differentiating into similar numbers of species. A vegetative mistake will cost no more to a chestnut than to a chickweed: it will merely prevent one seedling growing. But a sexual mistake will take toll only at maturity. That is a few weeks later in the weed but only many years later, after enormous biological capital has been locked up, in the tree. We should therefore expect trees to show greater stability of chromosome numbers.

What do we find? A survey of all the chromosome numbers of flowering plants in the *Chromosome Atlas* makes it possible to reach a general conclusion. The stability of chromosome numbers is correlated with the length of the reproductive cycle. Basic numbers often vary within species and usually vary within genera of annual plants. Among rapidly maturing perennials they usually vary within a tribe. But among slow-growing long-lived shrubs and trees there is a constancy which often transcends the limits of families. The gradation shown by the series of supposed dibasic polyploids in Table 13 is characteristic of the whole range of variation in flowering plants. As a rider to this generalization we may add that B chromosomes do not appear to exist in woody plants. They arise from, and in a sense represent, the extreme of instability and experimentalism in basic numbers. Thus we have the paradox that in their evolutionary stability the chromosomes are subordinate to the character of the organisms which they themselves determine.

Let us put these differences on to a time scale. In annual plants basic chromosome numbers are usually fixed only for a few thousand generations, often varying therefore within one species at one time. In woody plants, on the other hand, they are usually fixed for many millions of years. The rate of change, which, for the external properties of plants, is determined by reference to the fossil record is for the chromosomes determined by comparison of different contemporary groups. We can thus often infer that the present complements in many long-lived Angiosperms arose some time in the Eocene period; and in the Gymnosperms many complements have probably kept the same chromosome number since the Palaeozoic period. The consequences of this stability for the inference of descent we shall see later.

(iii) The Rejected Ones

Chromosome studies, as we have seen, point not only to the past but to the future. This capacity affects not only details and particulars: it affects the question of the scope of natural selection and indeed the whole future of evolutionary change. What proportion of living plants will contribute to the vegetation of future ages?

Comparison of the external forms succeeding one another in evolution makes it clear that only a small part of the species or even the families that once filled the world have left survivors today. Observation of the chromosomes of living plants tells us that the same will be true in the future. More than this, it tells us which of the living forms will survive. It tells us at two levels.

First, in regard to the changes themselves, as we saw, all plants select a small proportion of those occurring for their use.

Secondly, the woody plants reject practically all changes in structure and number of chromosomes. This rejection of chromosome change by woody plants is, no doubt, partly a rejection after the event. Partly, however, it is due to a genotypically controlled stability of the chromosomes. For small chromosomes are less easily broken by irradiation than large ones.

Thirdly, an enormous proportion of plants have chromosome complements and hence genetic systems marked with the signs, not of decay, but of that stereotyping which precludes adaptation and therefore precludes ultimate survival. The polyploids, the apomicts, the ring-formers or complex heterozygotes, have all sacrificed their enduring future to the temporary convenience of propagation. They have taken the cash and let the credit go. This is true, as we shall see, not only of species in nature but also of species in cultivation.

What is interesting about this inference of wholesale failure is both the agreement and also the disagreement between the external evidence and the chromosome evidence; for the chromosome or genetic types which will be eliminated are not distinct as such in external form. One cannot recognize polyploids, apomicts or complex heterozygotes by the naked eye. On the contrary, they represent almost entirely small subdivisions of genera or species with no other peculiarity than being a little more vegetative and a little less polliniferous than their primitive wholly sexual, diploid, and non-hybrid relatives. They represent therefore a wastage of diversity at another level and entirely additional to what the evidence of fossil forms and of experimental mutation had already given us. They mean that the proportion of forms which have a chance of posterity, which was always believed to be small, now becomes many times smaller. And the eliminating action of natural selection in directing change, which on external evidence is seen to be great, becomes many times greater.

(iv) Direction of Change

One of the great uncertainties in considering external evolutionary change is in determining the direction of change. What

is primitive and what is advanced are often a matter of speculation. In regard to the chromosomes we can, as a rule, give a
direction to their different kinds of change which is based on
experimental evidence. The definiteness of this direction is of
course due to the strict requirements of co-ordination of the
chromosomes. The evidence is summarized in Table 15.

So far as the evolution of genes themselves is concerned the
evidence has been discussed by Darlington and Mather. It does
not concern us here except that we should notice that two
systems are involved. Blocks of genes are prevented from
crossing-over by inversions of the segments of chromosome
which contain them. Such blocks, if they survive, inevitably
become complexes or super-genes on their own. They are shut
off from the freely crossing-over related genes in evolution in
the same way in which plants that have become obligatorily
apomictic or self-fertilizing are shut off from their freely interbreeding relatives.

Structural changes of chromosomes thus push the genes
along determined evolutionary paths just as genes push the
chromosomes along such paths. The evolutionary action is
reciprocal and interlocking. On the other hand, it is more
likely that the evolution of the major gene from the polygene
depends, not on such crude physical changes, but on the
rearrangements of chemical bonds by intra-genic mutation.

Polyploidy is the change of whose directional possibilities
and limitations we know most. All polyploids must ultimately
come from diploids. High polyploids, to be sure, can give low
polyploids, as in *Saccharum*, *Poa* or *Rubus*, by failure of fertilization. But in general, even here, the pressure of change is
probably upwards. Diploids very rarely, even in experiment,
come from tetraploids and when they do, they are unhappy by
comparison. Even when the tetraploid happens to be the 'type'
it must be supposed to have arisen from the diploid group in
which we now find its diploid 'variety'.

Another genetic change of whose direction we can often be
certain is that from hermaphroditism to dioecism. The change
has often been partly or wholly reproduced in experiment and
odd dioecious species are scattered through the hermaphrodite

flowering plants very much like species with other systems favouring cross-breeding. Moreover, the sex chromosome mechanisms underlying them differ from the long-established animal systems in their simplicity. Dioecious species when they are isolated are therefore clearly derived from hermaphrodites. When they occur in large groups, as in the *Bryophyta*, this is not so certain.

Two types of long-term genetic change may occur together in the same genus and reveal what sequences and combinations of events are in fact possible. It seems to be commoner in the flowering plants that dioecious species occur as isolated diploids, for example, in *Rumex, Humulus, Bryonia*. In *Empetrum* there is a diploid dioecious species in Europe which seems to have given rise to a tetraploid hermaphrodite form colonizing the arctic. Probably the tetraploid has dropped its ancestral dioecism which is no longer profitable.

On the other hand, in *Fragaria*, the diploid species are hermaphrodite and all the polyploid species ($4x$ in Asia, $6x$ in Europe, $8x$ in America) are partly or entirely dioecious. In this case the polyploidy seems to have given an advantage to dioecism. This may be because it broke down an incompatibility mechanism which was necessary in the diploid. Or it may be because they began as auto-polyploids. In general, we may expect that the introduction of polyploidy will be associated with a change of the breeding system.

Finally, the breakdown of the ordinary sexual breeding system to give complex heterozygotes, subsexual, apomictic, and other genetically aberrant forms must be irreversible except, as we saw in *Clarkia*, where the change is frustrated before it is complete. We have seen the stages by which these specialized systems arise and give a vast proliferation of forms like the 'species' of *Oenothera* and *Taraxacum*. We can also see them after they have declined leaving sometimes invariable, and isolated species like the pentaploid *Ochna serrulata* and sometimes what are called monotypic genera like *Rhoeo* and *Nicandra*.

In these instances we can show with Euclidean rigour the steps by which the genetic systems and the chromosome mechanisms have changed. We can point to the union of a diploid and a haploid gamete as the last sexual act in the ancestry of an apomict. We can also, with equal confidence, foretell its future elimination.

(v) The Inference of Descent

There are two kinds of inference about descent which we may make from what we know of the chromosomes. The first concerns the *future* of the individual or race or species; it can be derived only from the chromosomes and from other genetic evidence. The second concerns the *past* of the individual, race, species or larger group: it can be, and has been, largely derived from the evidence of general structure or phenotype. It is here that the chromosome evidence and the structural evidence, independent as they are in their origins, have to be compared. If we then find that they point in the same direction we need not pause for thought. But if, as sometimes happens, they point in different directions, we have to return to weigh the evidence with a more critical judgment.

How are we to sum up the weight and value of the chromosome evidence on the relationship of plants by descent in comparison with the structural or phenotypic evidence? The question reminds us of the comparison of the weight to be attached to generative and vegetative characters in the phenotype. The two comparisons, however, cannot themselves be compared. In assessing the chromosome evidence we have to apply quite a new rule, a differential rule. We have to say that for those with less stable chromosomes, usually the short-lived plants, the chromosomes have most weight below the genus; for those with more stable chromosomes, usually the long-lived plants, above the genus.

The immensely detailed variety of species formation is revealed in the short-lived plants by chromosome studies, and in no other way: but for the relationships of genera they are correspondingly and inevitably less important. The immense constancy of the chromosomes in the woody plants, on the other hand, overbalances the structural evidence on the remote relationships of these plants: but correspondingly and inevitably it throws little light on the relationships of species which stand massed by the hundreds and the thousands in groups of dead uniformity mitigated only by occasional polyploidy.

In the *Chromosome Atlas* it is not worth while doing more than to summarize the chromosome numbers of such unvarying genera. It is groups of genera, not of species, that provide the interesting comparisons.

This contrast is, of course, only a rough generalization. Its value is that it saves us wasting time by running unprofitable errands. The detailed systematics of many short-lived plants is completely fruitless unless combined with chromosome studies. It is not that the conclusions reached are contradicted or the names given are wrong: they merely prove to be useless conclusions and meaningless names. On the other hand, intensive chromosome study of species and varieties of a woody genus proves to be merely repetitive and equally unrewarding. What we need with the woody plants are comprehensive surveys of genera and families. Already a survey of our present knowledge in this field shows astonishing but plausible long-range connections.

We have seen what these connections are like in relation to the Agavaceae which in this respect fall in with the woody plants. By similar comparison we can now see how the changes in basic numbers of the woody plants are related to such phylogenetic schemes as that of Hutchinson (Fig. 40).

This diagram is misleading unless it is understood that the names of the families are taken to represent directly, not their external structures, but their chromosome numbers; and the relation of the two is just the problem to which we cannot give a uniform solution.

It is a temptation of course to say, on the lines of our discussion of secondary polyploidy, that one group such as the Magnoliaceae are descended from another such as the Lauraceae. This conclusion, however, is justified only in a chromosome sense. We cannot put contemporaries in a time sequence. Nor can we translate our evolutionary processes from the internal to the external side, or conversely. This inability, this negative principle, in itself teaches us much.

A plant which has an advanced form of flower may have a

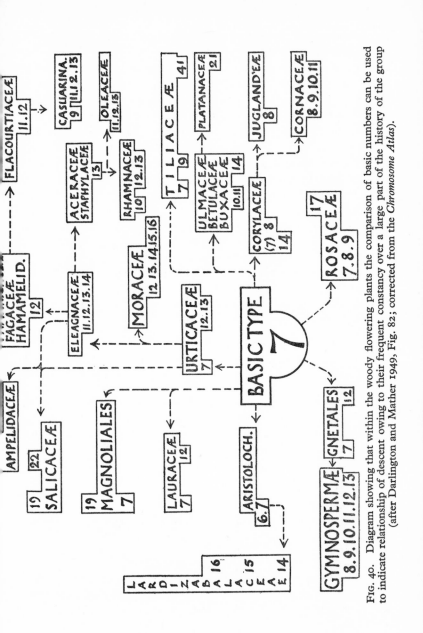

Fig. 40. Diagram showing that within the woody flowering plants the comparison of basic numbers can be used to indicate relationship of descent owing to their frequent constancy over a large part of the history of the group (after Darlington and Mather 1949, Fig. 82; corrected from the *Chromosome Atlas*).

primitive or prior chromosome organization (or, for that matter, pigmentation chemistry). The fact that flowers leave fossils does not make them more fundamental than chromosomes and pigments which do not. To be sure, in evolution the determinant and what is determined are continually interacting. And what is determined is of great practical importance in life. But the determiner comes first. Samuel Butler pointed out that a hen is an egg's way of making another egg. It is more precise (although it may be equally surprising) if we say that a tree is a chromosome's way of making more chromosomes.

Our diagram of chromosome phylogeny may therefore be of limited value by itself; but as part of the framework of evolutionary speculation and investigation it is indispensable.

(vi) The Origins of the Flowering Plants

In conclusion, what about the evolution of the higher plants as a whole? Clearly they owe their origin to the advantage that they gave by combining the wide dispersal both of pollen and seed with the mammalian-like gestation of the embryo in the seed by the mother plant.

Now we can see how the next step in evolution, the origin of the Angiosperms from Gymnosperm ancestors, so naturally followed. It was due to two connected inventions. The first was the long understood development of the style on the ovary. It acts as a sieve excluding whatever pollen is either too like or too unlike the mother plant in its genetic kinship and character. The second was the development by double fertilization of two embryos, the dummy endosperm serving as a buffer between the mother and the true embryo which it has to nourish. Thus, if we represent M as a maternal, and P as a paternal chromosome set, the nutritive channel is $MM \rightarrow MMP \rightarrow MP$. Thus each embryo is given an individually adapted *genetic environment* (Darlington 1971b).

By these accurate adjustments of the chromosome and reproductive mechanisms the Angiosperms have provided for the unprecedented range of possibilities in hybridization, dibasic polyploidy, subsexual reproduction, and so on, that we have seen; and all compatible with a fine reproductive economy. Hence we see the origin of the Angiosperms as a climax in the evolution of genetic systems surpassing even that of the mammals on the animal side.

INTERLUDE: POLYMORPHISM

WE HAVE now examined the first half of our problem. We have considered a number of respects in which the study of their chromosomes enables us to understand the character of natural species and the processes by which they have arisen in the past and are likely to arise in the future. But there is one general point which is more important than these details. It is that separate study, whether it is of ecology or distribution, variation in form or mode of breeding, chromosome number or cellular mechanism, always fails to tell us what is of most value. It is the connectedness between these different aspects of the life of plants and populations which reveals the meaning of evolutionary change.

In achieving this connectedness and this unity the traditional methods of systematics and nomenclature present us with certain obstacles. Our attention has to be directed, not to the earliest but to the latest studies, descriptions and names. Our assumption has to be, not of static but of dynamic species, species with changing populations of genotypes, with selective migration, with breakdowns of breeding systems, with a shattering of some groups, with the hybridization of others, with an interaction of genetic and external modes of isolation and sources of discontinuity.

Accordingly, therefore, we need to brush away the first artificial deceptions of systematics. Such are the notions of fixed species represented by fixed types with derived varieties. And the notion of a primitive species or group based on some arbitrary variable of subjective interest. These notions have nothing respectable about them but their antiquity.

Secondly we need to brush away the equally dangerous deceptions practised on us by nature. We have to learn that the solid mass of a species flowing continuously from generation to generation is an imaginative construction and also a misleading one. There is a continuity between generations seen in the chromosomes. But the plants seen by the naturalist or the experimental breeder are highly discontinuous. In following them we have to understand that each generation is derived from only one or two individuals out of a hundred in the preceding generation. We have then to master a double paradox: that it is derived from a similar proportion of one or two in a hundred of all preceding generations; hence its genes are derived from only one or two out of a million a few generations back. We therefore have to remember that within and behind the individuals there are genes and blocks of genes, processes and systems of processes, continually selected and mutually selected, saved or eliminated. Each of these things, the species, the plants, the genes, the processes and the relations between them, have their reality, their connections and above all their instability. All of them therefore contradict the permanent and retrospective conventions of classification and nomenclature. They allow us only to separate provisional classes and to give provisional names.

It was the view of Darwin, which still seemed to be true a century later to the Darwinian geneticists, Haldane, Fisher and Sewall Wright, that variation within a species was totally available for the construction of differences between species. This availability is still true of one half of the variation within species. Naturalists who ignore it are indeed wasting most of the interest of their subject. But the other half is also necessary for the understanding of the scheme of things. We now see it unmistakably and diagrammatically revealed by the chromosomes.

What appears in the behaviour of chromosomes as internal hybridity, and appears to the naturalist, in Ford's expression, as balanced polymorphism, and again appears in genetic analysis as the complex or the super-gene or, in its most extreme form as the sex chromosome system or the sex determining mechanism, all these are items or types of variation which are not available for creating the differences between species. They split species in the sense that they split the heredity of individuals. They even split populations of gametes and of zygotes. But they do not split whole genetic systems.

Again, take the B chromosomes. They are materials contributing to the variation within a species. But they do not contribute to splitting species. They do not contribute to splitting anything (Table 16).

TABLE 16

TYPES OF POLYMORPHIC SYSTEM (from Darlington 1971a)

Type	Notation	Function	Evolution	Incidence
1. ALLELIC or MENDELIAN (floating)	AA, Aa, aa	Maintains diversity or develops heterozygous advantage (HA) Restricts recombination	Single nucleotide to supergene With inversions, interchanges, duplications, fusions, heterochromatin	Man *Panaxia* *Drosophila* *Trillium* etc.
2. BACK-CROSSING (unstable)	(i) ss : Ss (ii) XX : XY	Enforces outbreeding: (i) by heterostyle incompatibility (HA) (ii) by diploid sex determination (HD)	Increasing divergence of differential segments with suppressed recombination	*Primula* *Lythrum* etc. All dioecious P. and A.
3. PERMANENT HETEROZYGOTE	(i) A + B : AB (ii) X + Y : XY	Maintains complex hybridity with self fertilization (HA) Maintains haploid sex differentiation	(i) Interchanges with balanced lethals (ii) Differential segments give haploid sex dimorphism	*Oenothera* *Isotoma* Bryophyta
4. MULTIPLE ALLELIC	$S_1, S_2, S_3 \ldots$	Enforces outbreeding by incompatibility (HA)	Accumulates differences by total heterozygosity near S locus	Higher and lower plants
5. PARTHENO-GENETIC (frozen)	ABC, abc	Stabilizes élite heterozygote (HA)	Heterozygosity for multiple inversions	*Drosophila mangabierai*
6. SUPER-NUMERARY	(i) $2x + 0, 1, 2, 3B \ldots$ (ii) $2x \rightleftharpoons 2x - 1$	Generates diversity by meiotic irregularity	(i) B chromosomes (ii) Isochromosomes	Short-lived P. and A. *Nicandra*

HA: Heterozygous advantage.   HD: Heterozygous disadvantage.

All polymorphisms, all permanent or internal hybridity, and the B chromosomes also in a different way, are indeed contributing, not to splitting species but to the opposite and equally important evolutionary task of holding species together. In the origin of two species from one we have complex or multiple differences which cannot be brought together without causing sterility. In polymorphism and internal hybridity and the B chromosomes we have complex differences which are brought together by breeding in every generation without causing sterility. Thus the hereditary or chromosome material is so organized that it can be split, as it were, in two different planes.

The differences and the similarities between these two types of split illuminate certain dark places in the theory of evolution. To give a simple example: for neither kind of split is any separation in space necessary; but, for the kind of split which gives rise to polymorphism and internal hybridity, separation in space, geographical isolation, is absolutely out of the question

Let us look now at a darker place. We said that one half of the variation in a species is available for splitting it and another half for binding it. But are these fractions equal to one another in real life? This is indeed one of the places in our view of nature, that no one has yet tried to explore. All that we can say is that a large part of variation is available for either purpose and is kept available by the breeding system until the moment when the advantage of one or the other becomes decisive. It is available in the many forms of what Ford describes as *transient polymorphism*.

There is now a new task we have to approach, a new connectedness that we have to examine. It is that between cultivated plants and the wild plants from which they have arisen. It affects only a few thousand species and what has happened to them over a few thousand years. But they are species which during this short time have been subjected to selection with such enormous effects, and to observation and experiment on such an enormous scale, that their understanding is of quite disproportionate importance.

The origins of cultivated from wild plants were first studied with profit by Darwin. He was concerned with using the evidence of the selection of cultivated plants to demonstrate the mode of evolution of wild plants. With the help of chromosome studies we are now in a position to reverse this process of inference. We are able to use our knowledge of the history of wild plants and of their variation and selection in nature to explain the origins of their cultivated relatives or descendants. This is the second half of our problem.

V. CULTIVATED PLANTS

1. WILD AND CULTIVATED PLANTS

THE key to our understanding of the evolution of living organisms is, as Darwin discovered, the study of what happens to them in the hands of man. During the last ten thousand years man has transformed the plants with which he has most concerned himself. The origin of cultivated plants, therefore, in some respects proved to be, as Darwin supposed, a model of the origin of species.

A joint attack on the problem of the origin of cultivated plants from the genetic and chromosome point of view has closely paralleled the results of the study of wild plants in two respects. It has revealed their history; and it has shown the diversity of this history, the great number of different ways in which the same result is achieved. Above all, it has shown the reasons for this diversity. Evolution is often dependent on hybridization; but it is also possible without any hybridization. It can even result from the prevention or suppression of a previous habit of crossing. Evolution is often dependent on polyploidy and other visible chromosome changes: but it is also possible without any visible chromosome change; that is, by gene change alone. Again, polyploidy is most important after the crossing of species; but, where fertility is not of crucial importance, it may produce results without any such hybridization. All these principles apply equally in wild and in cultivated plants.

The sources of variation, the principles of selection, and the conditions of geographical and genetic isolation, all operate equally in wild and in cultivated plants. Only one distinction may sometimes be made. It is that between what is natural and inferred and what is artificial and observed. But this distinction is, as we shall see, arbitrary and even in some cases fictitious. Most artificial selection of cultivated plants is natural to the extent of being unconscious, and most natural selection of wild species nowadays has an unsuspected artificial component.

Botany, the scientific study of plants, therefore depends on our directing our enquiries equally and jointly and with frequent exchange of ideas to the understanding of wild and of cultivated forms.

Within cultivated plants there is an endless variety of types according to their antiquity and mode of origin and improvement. When we discuss these differences one primary question presents itself: how long has the crop been improved? At one extreme there are the original grain crops, few in number and of a remote and undocumented origin. These are the plants to whose cultivation and improvement man himself owes the origin of settled society. At the other extreme are the ornamental plants, the frills and superfluities of civilization, plants existing in an enormous number of species, most of them little, but some much, improved, and that improvement being mostly recent and often very well documented.

Between the first and the last objects of cultivation there is a continuous series which we may put in historical succession, arranging the crops, as Burkill has done, according to their uses. One crop (as in *Brassica*) often to be sure serves as a source of greens, oil and fodder derived from leaf, seed and root. It is then successively selected for these different purposes. Indeed, we may suppose that the development of each crop created a demand for a necessary complementary element in diet and culture and forced man by expanding his population to the limits imposed by starvation into expanding his exploitation of the plants at his disposal at the same time that he expanded the area of territory he exploited nearly to the ends of the earth.

2. EARLY NOTIONS

Man's earliest belief was a simple one. He had no doubt that cultivated plants were an original gift of the gods. This belief may well have been weakened by the discovery of new and valuable crop plants existing exclusively in America, but another opinion almost as old and still widely held today, was ready to take its place. It was that cultivation itself directly improved the heredity of plants which were taken from the wild. Better soil, better fruit. And so the question was raised: what wild plants do they come from, and where are their wild ancestors to be found? Linnaeus tried to give an answer but there was more legend than history to guide him. It was not possible to say at that time whether maize and tobacco were Old or New World plants, and such distinctions as that between Indian and American cottons were unknown. So it was that in 1807 Alexander von Humboldt could still say that we knew nothing about the original sources of our most useful plants. Their origin seemed indeed 'un secret impénétrable'.

Our present elucidation of this mystery derives from three sources in the last century. The first is the work of De Candolle, his *Géographie Botanique Raisonnée* of 1855, and his *Origine des Plantes Cultivées* of 1882. De Candolle's methods and materials were new and enlarged. In detail he revised the earlier information by correlating it with the new evidence of archaeology and palaeontology, of history and linguistics. Ancient history and language, however, were not always helpful. As Pliny puts it: the kinds of grain are not the same everywhere and those that are the same have not the same names.[1] Nor were his premises and his purpose other than Linnaean: he wished to show, and did show to his own satisfaction, that the different kinds of cultivated plants could be traced back to wild ancestors represented by species still living and still recognizable. Of his 247 ancestral species, 199 Old World, 45 New World, and 3 uncertain, only 7 did he consider extinct.

This view, over-comforting by our present standards, was due to De Candolle's Linnaean simplicity. *Varietates laevissimas non curat botanicus*. Species were comprehensive units and their internal relationships could and should be neglected. The

[1] 'Frumenti genera non eadem ubique: neque ubi eadem sunt, iisdem nominibus. Vulgatissima *far*, quod *adoreum* veteres appellavere, *siligo*, *triticum*. Haec plurimis terris communia. *Arinca* Galliarum propria, copiosa et Italiae est. Aegypto autem ac Syriae, Ciliciaeque et Asiae, ac Graeciae peculiares *zea*, *olyra*, *typhe*' (*Hist. Nat.* XVIII, 8).

Pliny was referring to *Triticum monococcum, dicoccum, turgidum, durum,* and *vulgare*; but which was which he could not, and we cannot, truly tell. The situation for species then was as confused as it often is for varieties now.

wheats could best be divided into two species and certainly not more than four. All the cultivated races of *Brassica*, even swedes apparently, were still to be seen wild in Europe or Siberia: and worst of all was his opinion of the method of transformation. For such a change as that which, in his view, turned *Raphanus raphanistrum* into *Raphanus sativus* could arise by the direct and Lamarckian effect of cultivation.

Darwin, largely using the same evidence as De Candolle, takes a contrasted point of view. He wanted to conclude that large changes have taken place under selection. In his *Variation of Animals and Plants under Domestication*, published in 1868, Darwin was interested in showing that those connections with wild forms which De Candolle claimed to have established are no longer to be traced. The ancestral wild plants, he argued, were exceedingly poor. His own experience of primitive people suggested that in the beginning man was 'compelled by severe want to try as food almost everything which he could chew and swallow', a view with much biblical corroboration. He concluded that 'many of these plants have been profoundly modified by culture'. How they were modified was a question on which he changed his mind. In 1859 it was by selection; in 1868 (as we shall see later) he developed his theory of selection. But he also fell back on the assumption of a direct Lamarckian effect which he expounded under the new name of 'pangenesis'. His argument was that a direct effect of a changed environment in creating new variation would compensate for the loss of old variation by the 'blending inheritance' which he had taken to be a law of nature.

Darwin, however, did not claim to understand the 'laws of inheritance', and it was here that Mendel provided the decisive correction in his *Versuche über Pflanzen-Hybriden*, published in 1865. Mendel, like Darwin, was interested in cultivated plants because he believed they could tell us most clearly what we needed to know about heredity and variation in every kind of plant and animal. Excluding Lamarck he makes three points of which we still need to be reminded:

'No one will seriously maintain that in the open country the development of plants is ruled by other laws than in the garden bed. Here as there, changes of type must take place if the conditions of life be altered, and the species possesses the capacity of fitting itself to its new environment.'

Secondly, 'were the change in the conditions the sole cause of variability we might expect that those cultivated plants which are grown for centuries under almost identical conditions would again attain constancy. That, as is well known, is not the case.'

Thirdly, 'our cultivated plants are members of *various hybrid series*, whose further development in conformity with law is varied and interrupted by frequent crossings *inter se*. The circumstances must not be overlooked that cultivated plants are mostly grown in great numbers and close together, affording the most favourable conditions for reciprocal fertilization between the varieties present and the species itself.'

Thus it was to their special conditions of hydridization and selection operating under the same laws as in the wild, that Mendel attributed the development of cultivated plants. How these conditions have operated we are now in a position to see.

3. CENTRES OF DIVERSITY

The methods and theories of De Candolle, Darwin and Mendel were first combined in the work of Vavilov and his school. The Russian workers made a geographical survey of the genetic variation in the staple crop plants of the world. In their interpretation, following De Candolle and reversing Darwin, they applied the ideas derived from wild plants to the understanding of the cultivated plants. They claimed to have found that the species of cultivated plants showed centres of diversity characteristic of each. This diversity was true not only of the crop plants themselves but also of their parasites and enemies. On the first analysis (1926) Vavilov argued that six main centres could be named as sources of nearly all genetic variation. Later (1935) he increased the number to eleven or twelve. Further, he argued that the centre of diversity was the *centre of origin* of the cultivated 'species'.

This view can appear satisfactory in particular cases when the relatives and antecedents of a species of cultivated plant have largely disappeared. For we can then imagine that the cultivated species arose by one simple ameliorative process. But when we know more we find that species in cultivation no less than in nature arise and develop by utterly diverse processes, by

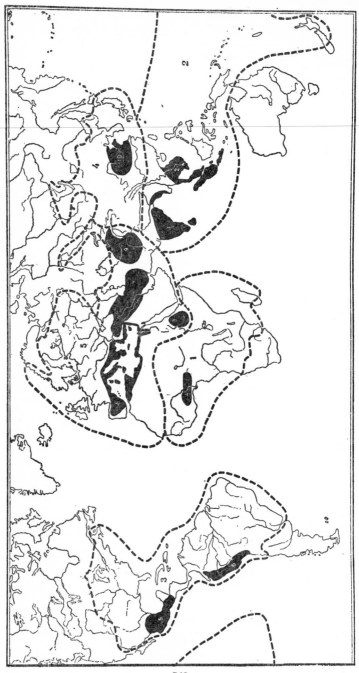

splitting, by shattering, by crossing, by doubling their chromosomes and so on. The results likewise are utterly diverse in their coherence or their lack of it. Some species therefore (whether in cultivation or in nature) can indeed be said to have had a single and sudden origin, localized and capable of being located. With others, however, the origin is no origin at all but a gradual transformation extending over wide areas and long periods and shifting its focus in the course of time. Between the two extremes there are gradations.

Whether sudden or gradual in its appearance, a cultivated or wild species will (if natural selection plays any part) come to show a local diversity related to its ecological diversity and this in turn will always be greatest in the mountain-and-valley regions nearest the equator. Here it was that Vavilov claimed to have found the expected diversity. If then we take Vavilov's centres of origin as centres of development his work may be treated as a starting point for our study as Kuptsov treats them (Fig. 41, Table 17).

A few examples will make this point clear. The archaeological evidence with the results of experimental breeding show that the new forms of bread wheats which are hexaploid arose in the seventh millenium B.C. from hybridization in Asia Minor between the tetraploid Emmer wheat which had been brought into cultivation in the Fertile Crescent and a diploid species of *Aegilops* growing wild in the same region. Its origin

---

FIG. 41. Historical map of the development of agriculture up to the great navigations. The chief primary regions of agriculture are shown in black:

1. Indian	2. Indonesian	3. Ethiopian
4. Nigerian	5. Mexican	6. Peruvian
7. Near Asiatic	8. Mediterranean	9. Central Asiatic
10. Chinese		

All of these, except 4, are derived from a neolithic stage. The main agricultural regions of the following period of expansion (with the approximately corresponding linguistic group names in brackets underneath) are numbered in reversed chronological order and separated with broken lines:

1. Negroid (Bantu)	2. Australoid (Hindu & Indonesian)	3. Americanoid
4. Mongoloid	5. Europoid (Aryan & Semitic)	(after Kuptsov 1955).

TABLE 17

TABLE 17

REGIONS OF ORIGIN OF CROP PLANTS (after Vavilov, revised in the light of work by Baker, Barrau, Burkill, Collins, Helbaek, Hutchinson, Kuptsov, Rick, Salaman, Simmonds, Whitaker, Zohary and others).

1. S.W. ASIA

Triticum diocccum, Emmer
Triticum vulgare, Bread Wheat
Hordeum sativum, Hulled Barley
Linum usitatissimum, Linseed
Lens esculenta, Lentil
Pisum sativum, Pea
Cicer arietinum, Chick Pea
Brassica campestris, Rape
Papaver somniferum, Opium Poppy
Cucumis melo, Melon
Daucus carota, Carrot
Ficus carica, Fig
Punica granatum, Pomegranate
Prunus avium, Cherry
Prunus amygdalus, Almond
Vitis vinifera, Grape Vine
Pistacia vera, Pistachio Nut
Diospyros lotos, Persimmon
Phoenix dactilifera, Date Palm
Crocus sativus, Saffron

2. MEDITERRANEAN

Vicia faba, Broad Bean
Brassica oleracea, Cabbage, etc.
Brassica napus, Swede Rape
Olea europaea, Olive
Ceratonia siliqua, Carob
Allium sativum, Garlic
Lactuca sativa, Lettuce
Beta maritima, Beet
Asparagus officinalis, Asparagus
Pastinaca sativa, Parsnip
Rheum officinale, Rhubarb
Humulus lupulus, Hop

3. EUROPE

Avena sativa, Oats
Secale cereale, Rye
Ribes spp., Currants
Rubus spp., Raspberries

4. ABYSSINIA

Eragostis abyssinica (for bread flour)

Eleusine coracana, Finger Millet
Dolichos lablab, Lablab Bean
Guizotia abyssinica, Niger Seed
Ricinus communis, Castor-oil Bean
Coffea arabica, Coffee
Hibiscus esculentus, Okra

5. CENTRAL AFRICA

Sorghum vulgare, Millet
Cola acuminata, Kola Nut
Sesamum indicum, Sesame

6. WEST AFRICA

Elaeis guineensis, Oil Palm
Ceiba pentandra, Kapok

7. CENTRAL ASIA

Panicum italicum, Millet
Fagopyrum esculentum, Buckwheat
Cannabis indica, Hemp
Phaseolus aureus, Mungo Bean
Brassica juncea, Indian Mustard
Spinacia oleracea, Spinach
Ocimum basilicum, Sweet Basil
Pyrus communis, Pear
Pyrus malus, Apple
Juglans regia, Walnut

5. INDO-BURMA

Oryza sativa, Rice
Cajanus cajan, Dhall
Phaseolus aconitifolius, Math Bean
Phaseolus calcaratus, Rice Bean
Dolichos biflorus, Horse Gram
Vigna sinensis, Asparagus Bean
Amaranthus paniculatus, Amaranth
Solanum melongena, Egg Plant
Raphanus caudatus, Rat's Tail Radish
Colocasia antiquorum, Taro Yam
Cucumis sativus, Cucumber
Mangifera indica, Mango
Gossypium arboreum, Tree Cotton, $2x$
Corchorus olitorius, Jute

TABLE 17 *(contd)*

Piper nigrum, Pepper
Acacia arabica, Gum Arabic
Indigofera tinctoria, Indigo

8. S.E. ASIA

Coix lachryma-jobi, Job's Tears
Dendrocalamus asper, Giant Bamboo
Dioscorea spp., Yam
Zingiber spp., Ginger
Citrus maxima, etc. Pomelo
Musa spp., Bananas
Cocos nucifera, Coconut
Elettria cardamomum, Cardamoms
Myristica fragrans, Nutmeg
Curcuma longa, Turmeric

9. CHINA

Aleurites moluccana, Tung Oil
Avena nuda, Naked Oat
Glycine max, Soya Bean
Stizolobium hasjoo, Velvet Bean
Phaseolus angularis, Adzuki Bean
Phyllostachys spp., Small Bamboos
Zizania latifolia, Manchurian Rice
Brassica chinensis, etc., Pak-choy
Allium fistulosum, etc., Welsh Onion
Raphanus sativus, Radish
Prunus armeniaca, Apricot
Prunus persica, Peach
Citrus nobilis, etc., Orange
Broussonetia sp., Paper Mulberry
Morus alba, White Mulberry
Camellia (Thea) sinensis, China Tea

10. MEXICO

Zea mays, Maize
Ipomoea batatas, Sweet Potato
Phaseolus vulgaris, Kidney Bean
Maranta arundinacea, Arrowroot
Capsicum annuum, etc., Red Pepper

Gossypium hirsutum, Upland Cotton
  (4x)
Agave sisalana, Sisal Hemp
Psidium guayava, Guava

11. U.S.A.

Helianthus annuus, Sunflower
Helianthus tuberosus, Jerusalem
  Artichoke

13. CENTRAL AMERICA

Cucurbita spp., Squash, Pumpkin,
  Gourd, etc.

14. PERU

Solanum tuberosum, Potato
Chenopodium quinoa, Quinoa
Phaseolus lunatus, Lima Bean
Canna edulis, Canna
Lycopersicum esculentum, Tomato
Gossypium barbadense, Sea Island
  Cotton (4x)
Carica papaya, etc., Papaya
Nicotiana tabacum, Tobacco
Cinchona calisaya, Quinine

15. CHILE

Bromus mango, Mango Grain
Madia sativa, Chilean Tarweed

16. BRAZIL-PARAGUAY

Manihot utilissima, Tapioca
Arachis hypogaea, Ground Nut
Phaseolus caracalla, Caracol
Theobroma cacao, Cocoa
Ananas comosus, Pineapple
Bertholletia excelsa, Brazil Nut
Anacardium occidentale, Cashew Nut
Passiflora edulis, Passion Fruit
Hevea brasiliensis, Para Rubber

NOTE ON SECTION 13

Cucurbita. The cultivated forms have baffled us in the past. Whitaker has established that they are wholly American in origin and all are diploids (2n = 40). One of them, *C. ficifolia,* misleadingly known as the Malabar Gourd, is a perennial confined to the highlands of Central America. It is easily distinguished from the other four which are lowland-living annuals. These four may be arranged by their probable origins in a geographical sequence from Texas to Peru as *C. pepo*, *C. mixta*, *C. moschata* and *C. maxima*.

The distinctions between these species are in the flower and the seed. There is a basic or common type of fruit in all four which is that of the thick-skinned Winter Squash or Pumpkin. But *C. pepo* has produced, both thin-skinned variants, the Summer Squash or Vegetable Marrow, and a profusion of thicker-skinned forms, the Ornamental Gourds. The first three species are all said to include Crookneck Gourds and the last, *C. maxima*, includes the Turban Gourds remarkable for having a unique inferior/superior ovary.

Presumably there has been parallel selection among products of crossing between geographical races of a species-complex which has now largely disappeared in the wild. Yet according to Whitaker their reciprocal crosses all fail to give F_1 hybrids. They can mix only through back-crosses. Probably combined cytoplasmic and nuclear differences have created sterility barriers between them. But these barriers, and the species they have created, have arisen under domestication. For all four species yield F_1 and F_2 hybrids when crossed with a wild species found in Central America, *C. lundelliana*. This species therefore seems to be close to the common ancestors of the whole group.

Some differentiation evidently existed however between ancestral forms at the beginnings of domestication. Already *C. pepo* seeds have been identified near Tamaulipas in Mexico in remains dated 3000 B.C. *C. moschata* and *C. ficifolia* have likewise been mentioned in the famous cave at Huaco Prèta in Peru dated 2000 B.C. Undoubtedly the double use of these fruits for food as well as for food vessels was well understood at the pre-pottery beginnings of settlement. It was also quickly understood by the discoverers who carried the American gourds to Asia, Europe, and Africa, where their American origin was unknown.

was therefore sudden and local and far to the west of Afghanistan where Vavilov found its greatest diversity.

Again the ancestors of the cultivated potato grows wild in Peru in immense variety. From here, however, according to Salaman, only a minute sample was brought to Europe in the sixteenth century as the basis of modern improvement.

There are other oddities that will not fit into any general scheme. The immensely polyploid Black Mulberry ($22x = 304$) is believed to come from Persia, but it has no diversity so that its origin cannot be deduced on Vavilovian lines. The Date Palm is not assigned by Vavilov to any of his regions, but it came from the Persian Gulf: it was brought to India, like tea and the sago palm, by the British East India Company. The Cocoanut Palm was assigned by Vavilov to the Malayan region. But its fruit, like certain gourds, being sea-borne, must be given the East Indian seas as its centre.

There are more general difficulties in Vavilov's simple scheme. The geographical distribution of wild plants has to be considered not merely, as Ridley did in his classical work, in relation to their means of dispersal. It has to be corrected by a knowledge of particular varieties which can now be distinguished by their chromosome numbers. For example we know that some species of *Ipomoea* are distributed by sea—notably *Ipomoea insularis*. But the hexaploidy of the sweet potato, *I. batatas*, both in Peru and in Polynesia makes its identification more certain. It leads to the outstanding contribution of botanical genetics to pre-history in demonstrating the spread of the sweet potato from Peru to Polynesia, a spread which proves the meeting of Old and New World Peoples assumed by Hornell and Heyerdahl in Easter Island some thousand or more years ago. It was later from Tahiti, as Burkill and Barrau have pointed out, that the Polynesians carried the sweet potato to the less equatorial regions where it became indispensable, namely Hawaii and New Zealand.

On the other hand a recent distribution by ocean currents before the intervention of man has been assumed by Burkill and by Merrill for *Lagenaria* the bottle gourd or calabash. This invaluable plant having floated across the Atlantic must have existed wild in both hemispheres and have been domesticated in South America perhaps even earlier than in Africa.

Similarly hemp grows wild over a vast region of Asia from the Caucasus to Mongolia and it is still, as we shall see from Vavilov's own account, continually being taken into cultivation.

There are other plants which owe their wide distribution to cultural rather than botanical considerations. The Mediterranean sea, for example, a barrier to wild plants, has been a means of dispersal and a bond of union for plants of established cultivation. It is à *propos* of this dispersal that we must consider, first, the origin and spread of agriculture and secondly, the strangely significant interactions that have occurred during the development of agriculture between three apparently separate processes: the selection, the cultivation, and the migration, of the plants that man took under his care. Or, shall we say, of the plants whose evolution became bound up with that of man.

147

4. ORIGINS OF AGRICULTURE

(i) The Evidence of Archaeology

Early sites of settled agriculture have now been excavated in a line around the Fertile Crescent with branches into the Persian and Anatolian tablelands. The discoveries in these areas, particularly of plant materials, show that the early cultivators were connected. Combined with the physical dating of deposits by the radio-carbon method, they have made it possible to reconstruct the steps by which agriculture arose.

The earliest settlements in the region are dated in the seventh millennium B.C. and lie at heights between 400 and 4000 feet. They reveal, according to Helbaek, Emmer Wheat and two-rowed barley, which are derived directly from local wild plants. They also reveal the hexaploid Bread Wheat which seems to have arisen in the earliest cultivation. Now the two wheats are richer grains than barley and they often support civilization alone which barley never does. Moreover wild barley, *Hordeum spontaneum*, is found widely from Central Asia to Morocco while the wild Emmer, the tetraploid *Triticum dicoccoides*, is found only between the upper Tigris and Palestine. It therefore seems likely that wheat was the critical crop, the key to the origin of settled cultivation.

By 5500 B.C. cultivation had spread over a wider area and the diploid Einkorn wheat was being grown, mixed (as we shall suggest, unintentionally mixed) with Emmer. A thousand years later cultivation had reached the swampy plain and irrigated cultivation had begun. Soon agriculture began to appear in Greece, in Egypt, and in the Indus valley. By 3000 B.C. it was spreading from the Danube valley into northern Europe. The origins of cultivation are not found till later in the third millennium in northern China, in southern India or in Nubia.

The timing of the development of cultivation in different regions is what one would expect on the assumption that the earliest cultivators, supported by their regular food supplies, had gradually multiplied and under pressure of crowding had moved into new regions in search of new land. But in doing so they had changed the character of some of their crops, they had lost others, and they had picked up many new ones. They had certainly begun with peas and lentils already in Kurdistan; and early they had also had flax which under irrigation became cultivated for its oil-seed as linseed. But in crossing Asia they had picked up new grains, millets and buckwheats, and in entering China they had acquired the soya bean and much later, coming from the south, rice.

While this was happening in the Old World, perhaps a little later and certainly more slowly, cultivation was also beginning in the New World. The critical event in America might seem to be the domestication of maize. Yet parallel developments in the third millennium B.C. show maize in the Bat Cave in New Mexico and people cultivating Lima beans, tubers, gourds, and cotton, but without maize, in Huaco Prèta in northern Peru. This absence of maize in a country where it was later to become the mainstay of agriculture suggests independence of the earlier and later developments. Yet an alternative explanation is that the maize in its poor original forms was not the foundation of agriculture either in Mexico or Peru. It was only in its vastly improved forms that it supported, along with beans and tubers, the later climax of civilization in both countries.

A point of importance here is the altogether extravagant process of evolution entailed by the development of maize. The change in its evolution (to which we shall return) seems to be the most profound ever produced by human selection. It is not surprising therefore that it occupied a longer period than the development of the Old World cereals.

The pre-historic and historical evidence may be summed up in a diagram showing the neolithic expansion which carried agriculture from its earliest centres but partly shed the plants it carried with it. This expansion is superimposed on the world-wide palæolithic expansion which had led to the occupation of the Americas and Australasia during the previous 20,000 years (Fig. 42).

What kind of correspondence do we find between the historical evidence and that derived from contemporary geography?

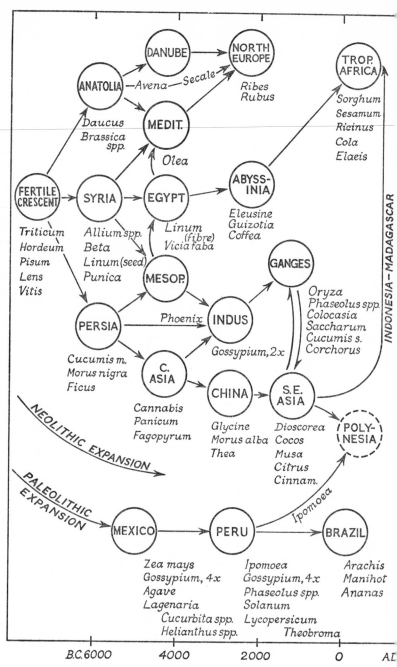

The geographical situation is summed up in Kuptsov's map which distinguishes between the ancient centres of cultivation and the large regions of commerce and exchange which developed round them in the period between the beginning of the bronze age, 3000 B.C., and the great navigations, A.D. 1500. These large regions are seen to have developed to a great extent in isolation from one another.

We must at once notice an apparent contradiction. On the one hand it is supposed that the movement of men spread cultivation over the whole of the Old World. On the other hand it is assumed that the movement ceased and was followed by the separation and isolation of great regions. This becomes less of a paradox when we recall that agriculture leads to settlement and also to the growth of dense populations supported by the canal and terrace cultivation built up by their own labours. These settled activities as they accumulate and by their very nature begin to impede the movements which gave rise to them. Moreover felling of trees and grazing of pastures by farmers developed in Asia, Arabia and Africa deserts which broke, or nearly broke, the connections that the farmers themselves had first made.

What then is the connection, we now have to ask, between the origins of agriculture in different parts of the world? It is beyond doubt that cultivation began independently in the Old and in the New World. Within the two hemispheres, on the other hand, connections evidently exist. The question is how crops changed in cultivation. And how did new crops come into cultivation especially when the cultivators themselves moved from one region to another. To these questions several answers are now provided by the processes of cultivation.

FIG. 42. Provisional diagram showing the primary expansion of cultivation by the movement of cultivators with their changing crops in the Old World from the ancient east, and independently in the New World from Mexico and Peru. The centres of redevelopment by new domestication are arranged in geographical relation and in chronological sequence. The arrangement depends on the correlation of archaeological dating with plant geography checked where possible by genetic tests, chromosome numbers and the Engelbrecht-Vavilov criteria. The discontinuity between regions is formal and arbitrary as in Fig. 41.

(ii) Crops from Weeds

Leaving the historical and geographical problem we must now consider the activities of the individual cultivator in relation to the particular crop. The first question here is that of how man in the beginning took up his crop plants and came to change them. The assumption made by Darwin and De Candolle, was that he deliberately chose his crops and then by selective breeding changed them. A radically new approach was made to the question in 1916 by Engelbrecht, a German plant geographer.

TABLE 18

ORIGINS OF CROPS FROM WEEDS[1]

PRIMARY CROP	SECONDARY CROP	ORIGIN AS WEED
Triticum vulgare	Secale cereale	S.W. Asia
Hordeum vulgare and		
Triticum dicoccum	Avena sativa	Europe and W. Asia
	Eruca sativa	C. Asia
	Camelina sativa	Transcaucasia
Linum usitatissimum	Spergula linicola	Transcaucasia
	Brassica campestris	Transcaucasia
Fagopyrum esculentum	F. tataricum	Altai
	Vicia sativa, etc.	S.W. Asia
Cereals	Pisum arvense	
	Coriandrum sativum	Transcaucasia
	Cephalaria syriaca	Asia Minor
Various crops	Cucumis trigonus	Turkestan
	Abutilon avicennae	Med.—China

[1] Derived from the observations of Engelbrecht 1916 and Vavilov 1926; cf. Schiemann 1943.

Engelbrecht's view was that there were certain primary crops which had offered themselves to the earliest collecting people by growing near their temporary settlements. These were what he called *habitation weeds*. They were favoured by the animal debris around dwellings because they had high nutritive requirements. They began, of course, as hardy wild plants usually with a long growing period.

The first of these plants were valued for carbohydrate like wheat, millet (*Sorghum* and *Setaria*), beans (*Vicia faba*), maize, sugar cane and potato. Later oil seeds were valued, linseed, hemp and rape. Others which followed were gourds, tomatoes and carrots. These plants, Engelbrecht argued, sought man out as much as he sought them out. Like the domesticated animals, they met him half way.

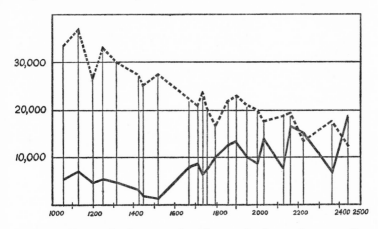

FIG. 43. Changes in the proportion of rye and wheat, when ascending from the lower to the upper zones of mountainous Zeravshan. Abscissae: altitudes in metres. Ordinates: average number of grains per Kg. Continuous line: rye, broken line: winter wheat.

As we ascend the mountains the amount of rye in the samples increases. The supplanting of wheat by rye is especially intense at altitudes above 1600 m. and beyond 2200 m. rye begins to prevail (Batyrenko 1926).

Following these primary crops were others which forced themselves on the cultivated fields. They were not habitation weeds but *crop weeds*. The evidence that this is an effective mode of origin of a new crop was provided by the two buckwheats. In Germany *Fagopyrum tataricum* grows as a weed in crops of *F. esculentum*. But in India as cultivation rises above the level of 10,000 feet the hardy *F. tataricum* begins to replace more productive *F. esculentum*. It becomes, according to Engelbrecht, a crop in its own right.

Similarly Vavilov and his school found that, as conditions became colder and less favourable for the primary crop, rye displaced wheat in cultivation (Fig. 43). If this can happen as cultivation nowadays moves into the mountains in Central Asia it could also have happened as cultivation moved north in Europe 4000 years ago. Accordingly Vavilov supposed that rye came into cultivation as an impurity. When *Triticum dicoccum* was shipped from Athens to the Scythians on the Dnieper in the fifth century B.C., he suggested, this wheat would contain *Secale montanum* as a weed in the crop and as an impurity in the grain. What was a weed in Greece would then soon become a crop in Russia.

Thus it is, we may suppose, that hardier and less demanding, and quicker ripening weeds would have displaced the primary crop as man has extended his cultivation into less favourable country with colder or hotter, dryer or wetter climates and poorer soils. The precise sequence of events and relation of species must, of course, often be doubtful. For example, we cannot be certain, without more archaeological evidence, whether barley and *T. monococcum* were at first weeds of Emmer. *T. monococcum* seems to have been secondary; but it has existed in Anatolia as a pure crop since lower levels in Troy. Similarly we cannot be sure whether oats was harboured first as a weed of Emmer or of horse-beans. It certainly spread when agriculture reached western Europe. In a different way the leguminous fodder plants would establish themselves in the wheat stubble. And eventually they have taken their own place, sown under the wheat in a rotation.

Perhaps the most remarkable of these situations concerns the annual teosinte, *Euchlaena mexicana*. Maize we suppose to have been derived by selection from an ancestor very like this plant. But teosinte now appears as a weed in maize fields. And from this weed a new teosinte has arisen of which varieties are today cultivated. A whole sequence of crops, Engelbrecht suggested, might follow one another into cultivation in this way from a beginning as weeds. Such a sequence was, for example, *Oryza-Eleusine-Panicum-Paspalum*.

(iii) Unconscious Selection

We can now see connections between Engelbrecht's idea of the origins of cultivation and other apparently remote but

quite fundamental evolutionary notions. These concern in
the first place Darwin's theory of unconscious selection.

Engelbrecht and Vavilov did not refer to Darwin's views.
We should, however, try to remember the origins of this idea
for it is likely to play a great part in our understanding of the
evolution not only of agriculture but also of civilization.
Darwin's theory of unconscious selection was developed in his
Animals and Plants under Domestication (chap. 20). Here he
notes the extremes of *natural selection*, which is independent of
man, and *methodical selection*, which is systematically and
deliberately guided by man towards a 'predetermined standard'.
With them he contrasts an intermediate situation. This is how
he describes it:

'Unconscious selection is that which follows from men naturally preserving
the most valued and destroying the less valued individuals, without any
thought of altering the breed; and undoubtedly this process slowly works
great changes. Unconscious selection graduates into methodical, and only
extreme cases can be distinctly separated; for he who preserves a useful or
perfect animal will generally breed from it with the hope of getting offspring
of the same character; but as long as he has not a predetermined purpose to
improve the breed, he may be said to be selecting unconsciously.'

Darwin adds a footnote: 'The term unconscious selection has been
objected to as a contradiction; but see some excellent observations on this
head by Professor Huxley . . . who remarks that when the wind heaps up
sand-dunes, it sifts and unconsciously selects from the gravel on the beach
grains of sand of equal size.'

It will be seen that Engelbrecht's idea is closely related to
Darwin's. What the precise relationship is becomes clear when
we consider a second fundamental problem, that of the genetic
character of the seed sown in primitive agriculture.

The seeds of all crops when harvested for sowing contain
two kinds of impurities which in primitive cultivation form a
large part of the sowing. Indeed they reach equilibrium with
the main or primary crop. This situation must go back to the
beginnings of agriculture. In relation to the views of Engelbrecht and Vavilov these impurities are of three kinds:

(i) *Unconditional Weeds*, that is species of plants whose
presence is wholly deleterious to the crop and useless
to man: they have little to do with the evolution of the
crop.

(ii) *Conditional Weeds*, that is species not originally cultivated but which might by improvement, or in a second season, or for another purpose, or under changed environmental conditions, come to contribute to the crop.

(iii) *Variants* of the crop species itself, hybridizing with the crop and existing therefore, not like the first two, merely in ecological equilibrium with the crop, but as part of the genetic equilibrium of the crop population itself. These variants arise from two sources: internal recombination and hybridization with wild populations in the neighbourhood.

The two latter impurities are the basis of unconscious selection: the one leads to evolution as between crops, the other to evolution within crops. The evolution between crops is controlled by both migration and cultivation. The evolution within crops is especially controlled by cultivation. These two modes of evolution we shall now consider.

5. EVOLUTION BY MIGRATION

The first effect of unconscious selection on mixed crops in migration is, as we have seen, to change the species so that different regions come to have widely different crop species. The second effect, which partly competes with the first, is to change the character of the species. Kuptsov's five agricultural regions are the regions between which crop species changed in the early expansion of cultivation; within these regions it was rather the internal character of the species which changed.

The reason for the barriers between the great regions is the absence of movement of people. But beyond this is another double reason. The crop is dependent on people to carry it and the people are dependent on the crop to support them. Both are conservative and there is a reciprocal relation between them. The Greeks were thus limited by the olive as Islam is limited by the date palm and the Chinese by rice or soya. Already in the bronze age the fibres used for textiles divided the world: silk in China, cotton in India, linen in the west.

This conservatism of people in regard to the crops they grew

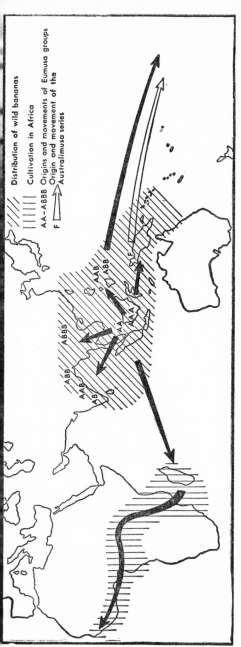

FIG. 44. Map showing the movement of the two sections of edible bananas since their cultivation began in the Indonesian region in the 1st millennium B.C.

(i) *Australimusa* (2n = 20: white arrow) the Féi bananas, related to *Musa maclayi*, consist of 20 clones which spread from New Guinea reaching Samoa, the Marquesas and Hawaii. Having failed to produce polyploids they seem to have been largely displaced by (ii).

(ii) *Eumusa* (2n = 22, black arrows) with two species *M. acuminata* (A) and *balbisiana* (B) whose chromosomes were variously combined to form 130 well known clones in S.E. Asia, largely polyploid. The extensions into Polynesia was from the Philippines. The two sections of banana are both represented *outside* the Indonesian region by identifiable seedless clones which arose from seedlings *inside* that region. They thus mark the movements of peoples. They confirm the double origins of the Polynesians and indicate the direction of migration across Africa.*

* (Simmonds 1962, *The Evolution of the Bananas*, reproduced by permission of Longmans, Green and Co., Ltd.)

157

was broken down in ancient times only by the conquest and subsequent hybridization of the people themselves. From this point of view the successive empires of the Persians and the Greeks, of Rome and of Islam were means of disseminating the crops grown in the ancient east; the first three brought fruit and vegetables into Europe; the last brought new crops and stock into Africa and also led to the development of native species.

These great movements not only shifted the crop; they changed it. Any wild species which survives alongside its cultivated forms will cross with them. The effects of this hybridization have been most clearly noted for *Triticum* in Syria and for *Lycopersicum* in Peru. Often the wild relatives will be better adapted to local conditions. *Beta maritima* and *Brassica oleracea*, like the other European vegetables *Crambe maritima* and *Asparagus*, are seashore plants first domesticated on the Mediterranean coasts. The cultivated forms of *Beta* and *Brassica* were carried through Europe and selected for many different purposes at the same time that they were crossing with wild forms. Each therefore diversified to produce many different crops in origin local and now universal. So too lettuces and rhubarb, carrots and turnips, cherries and currants, have been diversified and improved as they have passed through regions of Europe which were not the origin of any of them.

In Africa the movement has been in reciprocal directions. Canoe navigators brought the Indonesian crops to Madagascar and East Africa. Whence, as Simmonds has shown, carried across the continent, they set on foot the Bantu expansion (Fig. 45). Later the Arabs in pursuit of slaves in tropical Africa also, according to Burkill, collected there such important crops as sesame and sorghum.

Thus *Sesamum indicum* comes not from Afghanistan, as Vavilov had argued, but from Africa. It made its way through the 'Sabaean lane', that is, across southern Arabia to India and later China in the course of a millennium. Similarly, the sorghums to the north and south of the African rain forest united in the east where they were improved and were later, between A.D. 700 and 1100, carried by the Yemenite Arabs from

Zanzibar to India (Fig. 46). Rice came to India from the east while wheat, as we all agree, came from Persia. The present distributions of the three great cereals within India still reflect, on this view, their three directions of entry.

Fig. 45. Map showing the agricultural regions where the two groups of African sorghum millets grow, the region where they met, and the route by which Arab traders probably took them to India (from Burkill 1953).

There are also world travellers. Such is that ancient forage crop known to botanists by a Greek name as the Medish grass, *Medicago,* to Europeans by their own place Lucerne, and to Americans by the Arabic-Persian-Spanish name of Alfalfa or horse-fodder. This plant probably arose as an allo-tetraploid from diploid hybrids in the upper Euphrates region and was preserved chiefly by cultivation. It was brought eastward to

China in the second century B.C. and westward from Persia to Greece under the Persian Empire of Darius; still further to Switzerland and Spain under the Roman Empire; and under the Spanish Empire it was taken to the New World entering by way of Peru and Chile. Only later, as Hendry has shown, did it invade the United States. In the eighteenth century on the eastern side it came from England, where it had existed as lucerne for two centuries; and with much greater success in the nineteenth century it entered on the western side from Peru, where it had existed as alfalfa for three centuries. In the first part of this vast peregrination *Medicago sativa* had no doubt crossed with, and been replenished by, both diploid and tetraploid wild relatives but it had remained itself wholly tetraploid. Hence today, the agriculturally useful derived types exist wholly at the tetraploid level.

An even more complicated history is told of the many-named earth nut, ground nut or pea nut, *Arachis hypogea*, also a tetraploid. From Brazil it went to Peru long before the conquest. Thence it was carried to the Old World, to Africa and India, by the Portuguese, and to the Philippines by the Spaniards. In all these places it became readapted and specialized and from all of them returned again, with the slaves and after them, to Tropical America and the United States where its new variants have been recombined and their progeny reselected.

Migrational types of development are thus historically unquestionable. Their genetic significance is also unquestionable. Movement has a snowball effect on variation, the snowball being the variants that are collected by hybridization and selection on the track of the migrant. Later we shall see just what these variants are. We must remember, therefore that in all these enquiries our species names are true as representing not the fixed types known to systematics but rather the developing and accumulating systems of diversity known to genetics. This situation has not the formal simplicity with which Vavilov endowed it. His centres of diversity have sometimes been centres of origin, sometimes centres of selection, of migration of hybridization, or merely of cultivation. This last we can now consider.

6. EVOLUTION BY CULTIVATION

(i) Operational Selection

As soon as a crop is cultivated regularly, that is year after year, the processes of unconscious selection begins to work on the variation which is available within the species. This particular class of unconscious selection, which we may call operational selection, falls under three headings:

(*a*) *Tillage conditions* lead to the selection of larger forms, some of them polyploids, which can compete favourably with earlier forms on arable land with higher nutrition. This effect must be very rapid with certain crops which, on ploughed and fertilized and weeded land will compete favourably with the original wild types.

(*b*) *Sowing conditions* lead to the selection of forms with even and rapid germination. A slow and widely variable speed of germination favours survival in nature, as we have earlier seen in relation to *Nicandra*. The effect of this selection appears especially from comparison of cultivated and wild forms of hemp and of various Cruciferae and Solanaceae. The digestion of the tortoise remedies this defect in the wild tomatoes of the Galapagos, but such devices have to be dispensed with in treating cultivated seed. There is thus a change of selection. Sowing conditions also lead to selection of forms lacking the special mechanisms of natural seed distribution and implantation, such as the awns of grasses, which have no further use in cultivation and indeed become a disadvantage.

(*c*) *Harvesting conditions* lead to the selection of forms with non-dehiscent fruits, as in hemp, flax, lettuce, and the opium poppy. In the cereals this often involves two changes: a toughening of the rachis, and a loosening of the grain from the glumes or hull, so that it escapes less easily in the field but more easily in threshing. Where, as in *Aegilops speltoides*, one of the diploid ancestors of the polyploid wheats, Zohary has found two kinds of seed dispersal, such a balanced polymorphism is bound to be suppressed.

These processes were dominant in a neolithic stage of

cultivation. The climax of their achievement is seen in maize where the glumes and pales have been almost lost and protection of the grain has been undertaken by leaves from lower down the rachis which cover the whole ear. Here the steps of improvement have been lost. In the wheats however they have been preserved. The *Triticum* species from which these grain crops are derived fall into three groups according to their chromosome numbers, 14, 28, or 42, the last having been preserved only by cultivation. The stages of selection can be represented by a simplified arrangement of the species (Table 19).

Historical, geographical and chromosome evidence do not require that the development of wheats has been from one species now living to another. The species remaining merely show us the stages we expect to find, and we find these stages

TABLE 19

EVOLUTION OF THE WHEATS (*Triticum*)[1]

Series	WILD TYPE: Brittle Rachis, tough awned Glumes	FIRST STEP: Rachis less brittle	SECOND STEP: Rachis tough, Glumes loose and often awnless
Diploid $2x = 14$	aegilopoides (Asia Minor)	monococcum (relic)	—
Tetraploid $4x = 28$	dicoccoides (Syria)	dicoccum (relic)	turgidum polonicum durum (Mediterranean)
Hexaploid $6x = 42$	—	spelta (relic in the Rhine Valley)	sphaerococcum compactum vulgare = aestivum (Asia Minor)

[1] The earliest wheat discovered at Chatal Hüyük about 7000 B.C. was *T. dicoccum*. It must have arisen by hybridization of the wild form of *T. monococcum* with a species of *Aegilops* before the beginning of cultivation. *T. dicoccum* diversified into all the tetraploid forms in cultivation, *durum*, *turgidum*, etc., at the end of the Neolithic Age, *c.* 2000 B.C. *T. aestivum* is believed to have been present from near the beginning and to have given rise to the hexaploid complex of varieties or species in the course of migration.

in each of the chromosome series except that, as we might expect, the highest stage is not found in the diploid, nor the lowest stage in the hexaploid: they have never existed.

(*d*) *Mode of propagation.* The cereals, pulses and oil-seed plants are propagated only by their seeds and grown only for their seeds. Seed fertility is thus the be-all and end-all of their selection and evolution. It is not surprising therefore that they are all diploids or have achieved, by one of the mechanisms we noticed earlier, a regular pairing at meiosis without any abnormalities of chromosome number, association or movement.

The same is true of all crops which are grown for seed even though, like coffee and cocoa, they may be propagated vegetatively. The same is true also of cultivated plants grown for their fruits where their fruit depends on the setting of a single seed for its development. And it is even true wherever seeds are used for propagation regardless of their utilization. For in all these plants we not only need many seeds but we also need true-breeding seeds. Thus first unconscious and then conscious selection for seed fertility dominates the evolution of these crops and maintains functional diploidy.

This limitation, this absolute requirement of seed fertility, true breeding and functional diploidy, is removed only in crops where both the propagation and the utilization are divorced from seed production. This is manifestly the case in plants vegetatively propagated and vegetatively utilized like tea or potatoes. But it is also true of plants grown for their flowers, that is ornamental plants. And it is true of plants grown for their fruits where the fruit is many-seeded. Indeed it is noticeably advantageous for the apple and the pear with ten-seeded fruits to be triploid. The seed fertility is then curtailed to the right point for limiting the fruit formation and yielding fruits of the largest size with even crops every year. It is also advantageous for the banana to be triploid or otherwise altogether sterile for here the formation of seed is a positive disaster. In consequence the object of selection has been what we recognize as complete seed sterility combined with parthenocarpy, the capacity of the fruit to develop without the development of seed (Table 20).

TABLE 20

PROPAGATION AND UTILIZATION OF CROPS IN RELATION TO POLYPLOIDY

Propagation of Crop	Utilization of crop				
	Seed	Fruit		Flower	Leaf, Stem, Root
		1–2 seeds	many seeds		
Seed	Cereals Pulses Oil seed	Cocoanut Oil and Date Palms	Tomato	Annual Flowers	Green Vegetables *Nicotiana*
L.S.R.	Cocoa Coffee	*Prunus* spp Olive	*Pyrus* 2x, 3x *Vitis* 2x, 3x, 4x[1] *Musa* 3x, 4x[1] *Ananas* 3x[1] *Citrus* 2x, 3x, 4x[1] *Fragaria* 8x	*Prunus* 2x, 3x *Rosa* 2x, 3x, 4x *Dahlia* 8x *Tulipa* 2x, 3x *Narcissus* 2x, 3x, 4x *Iris* 3x, 4x, 5x	*Thea* 2x, 3x *Dioscorea* 6x, 14x *Ipomoea* 6x *Colocasia* 2x, 3x, *Saccharum* 14x
	all functional diploids no triploids	triploids and autopolyploids only with L.S.R. propagation			

[1] Many forms of *Vitis*, *Citrus*, *Ananas* and all edible forms of *Musa* have been selected for steril combined with parthenocarpy (Simmonds 1962).

(ii) Functional Diploidy

One of the great problems for the developing seed-propagated or seed-utilized crop is evidently how to preserve its seed production while making use of the opportunities offered by polyploidy. In this respect the great seed crops fall into three classes (Table 21). Diploid plants, notably maize and rice, have usually managed to become staple grain or seed crops of the world without increase of chromosome number. A few, such as linseed and sorghum millet, had already been polyploid in nature. A few others, however, notably bread wheat, have obviously depended on the change to polyploidy for reaching their supreme importance. How have they done it? The critical evidence has been obtained in the hexaploid bread wheat.

The experiments of Sears in Missouri, Unrau in Alberta, and Riley in Cambridge have made the origin of bread wheat and the relations of its chromosomes fairly clear. The species is evidently derived from crossing successively three diploid species such as are still growing wild in the Fertile Crescent. This crossing has been followed by a doubling of chromosome numbers as assumed from the discovery of its polyploidy by Sakamura in 1918. The first cross gave the wild tetraploid *T. dicoccoides* before the beginning of cultivation; the second gave the hexaploid *T. aestivum* soon afterwards.

TABLE 21

STAPLE ANNUAL SEED CROPS CLASSIFIED ACCORDING TO POLYPLOIDY AND MODE OF ORIGIN (after Darlington and Wylie 1955)

Diploids with a diploid origin	Polyploids with a polyploid origin	Diploids which have become polyploid
1. CEREALS		
Zea mays	*Triticum dicoccum*, etc.	*Triticum vulgare*
Oryza sativa	*Sorghum vulgare*	*Brassica napus* (?)
Hordeum vulgare	*Avena sativa*	
Secale cereale	*Panicum miliaceum*	
Fagopyrum sativum		
2. OILS AND FIBRES		
Gossypium spp. (O.W.)	*Linum usitatissimum*	none
Cannabis sativa	*Gossypium* spp. (N.W.)	
Brassica campestris		
3. PULSES		
Pisum sativum	none	*Arachis hypogaea*(?)
Vicia faba		
Cicer arietimum		
Lens esculenta		
Vigna unguiculata		
Glycine max		
Dolichos lablab		
Phaseolus spp.		

The three sets of 7 chromosomes in the hexaploid evidently correspond closely in structure and action. But their capacities for pairing at meiosis are somehow restrained. They do not form multivalents at meiosis, and the hexaploid behaves as though it were a diploid. The three sets are usually known as A, B and D and they pair at meiosis A-A, B-B, and D-D. Even in the haploid derived by parthenogenesis the 21 chromosomes form only one or two bivalents and these have only one chiasma apiece instead of two or three in the normal pairing. Further in hybrids with diploid species of *Aegilops* or *Secale* only two or three bivalents are formed.

Evidently the pairing at meiosis has become restricted by

some special mechanism. Its origin has been shown by Sears and Riley who have tested the 21 possible types of $6x - 1$ plants, so-called monosomics. They find that the absence of a particular chromosome B5 or more precisely its long arm, destroys the controlling device necessary for regular meiosis in polyploid cereals: the A, B and D chromosomes pair and form chiasmata and multivalents appear at meiosis. Fertility and true-breeding are thus lost together.

It is this multivalent formation which has been eliminated in the selection for fertility and for functional diploidy in the hexaploid and indeed in the tetraploids long before their cultivation began. To put it another way, the effect of selection for fertility in polyploid wheat has been to shift the mechanism of meiosis towards regular bivalent formation and segregation. These changes were due to specific modification of chromosome B5 which altered the control of meiosis. The means of control is evidently such as to prevent any of the changes of partner amongst homologues at the pachytene stage which occur in the absence of B5.

The selection process, revealed in this instance in a cultivated cereal, is probably not in any way peculiar to cultivation. Rather it is of the same character as what happens in all wild allo-polyploid species where high seed fertility is needed and where functional diploidy is achieved. Consequently it applies equally well to cultivated species like *Nicotiana tabacum* and *Gossypium barbadense* which may well have been tetraploid before they were taken into cultivation. The cultivated bread wheat thus reveals first how the high pressure of selection for seed fertility restrict the development of polyploidy in all annual plants; and secondly it reveals how they overcome the difficulty. The absence of polyploids in the pulses may be due to their failure to make the necessary change in the control of meiosis.

(iii) Loss of Ancestors

Yet another consequence of the natural process of cultivation which is to be expected on our view is that the cultivated forms of a species should sometimes have extinguished their wild progenitors. Their rapid development during the last 10,000 years has coincided with a vast change in the ecological conditions and the genetic and geographical relationships of their wild ancestors and indeed of the whole flora. Changes of climate have no doubt also played their part. The expanding desert might have, as it were, driven the ancestors of our bean, *Vicia faba*, into the irrigated valleys and into cultivation; but it has left survivors discovered by Zohary in Palestine.

The extinction of wild ancestors is, however, another way of saying the occurrence of complete domestication. Such a change may well have overtaken a number of species: *Lens esculenta*, *Allium cepa*, *Zingiber officinale*, and the date palm *Phoenix dactylifera*. It is even perhaps true of two ornamental species, both recent in cultivation, both diploids, both apparently free from suspicion of hybridization: the garden tulip and the Chinese primrose.

Quite another ground for the apparent loss of ancestors is the complete transformation of a species. There is nothing remarkable about the production of a new species in cultivation by polyploidy which is undoubtedly the origin of *Triticum vulgare* and *Dahlia variabilis*. But can such a change take place in a diploid and without crossing?

Zea mays is a diploid. Yet it has no wild progenitor or group of progenitors attributed to the same genus. The crop plant may have destroyed its wild relatives by cross-pollination which would unfit them for survival in the wild. But the tremendous range of variation of this species under cultivation leaves us in no doubt that it arose, as we saw, by selection from equally and unfailingly monoecious *Euchlaena mexicana* (Gallinat 1971).

The transformation of an ancestral pod corn with small ears like those of teosinte and grains and brittle rachis to give the hundred-fold heavier many-rowed ears with large naked grains of modern cultivated maize would seem to be the outstanding example of what selection can do. Superficially it would seem that the ancestral pod corn and its modern descendant deserve to stand in different genera. The experimental and chromosome evidence, however, both contradict the judgment of the naked eye in the following way. Common forms of teosinte produce fertile hybrids with maize. They have the same ten pairs of chromosomes, largely corresponding in form, and the hybrids show regular pairing. Indeed different forms of teosinte differ more in their chromosomes and produce less fertile hybrids with one another. The breeding and chromosome evidence therefore agree in suggesting that *Euchlanea mexicana* and *Zea mays* could be described as one species. The drastic external transformation of maize appears to break its connection with its ancestors. But it is merely concealing the genetic reality of an evolution which our weaker brethren are too timid to admit.

(iv) Breeding Systems

In all directions we find natural selection working on cultivated plants in their artificial conditions, working untended and, in its slow and secular effects, unobserved by man. But, before selection can act, variation must be present for it to act on. There is yet another set of conditions which has affected the origin of variation in cultivated no less than in wild plants. These are the conditions of fertilization referred to by Mendel. Here again our genetic understanding of the breeding system reveals the accidental, but none the less regular, effects of cultivation. The changes of habitat it brings with it are liable to change the breeding habits of plants in the following ways:

(*a*) *Inbreeding replaces Outbreeding.* The absence of pollinating insects in a new environment leads a normally crossing plant to fertilize itself. Tomatoes which can and do cross freely in their native tropics (cultivated forms even crossing with their wild relatives) are compelled to self-fertilize themselves in Europe. In these circumstances any change that assists self-fertilization is itself favoured by selection. Sometimes genetically self-fertile variations are selected in self-sterile species, e.g. in the plums. Sometimes variations of flower-shape are selected which favour self-fertilization, e.g. a short style in tomatoes and black-currants. And again in hetero-styled plants the slight capacity of the pin-type to fertilize itself is developed by selection, e.g. in *Primula sinensis*, in the 120 years since it was introduced into England (Mather and De Winton, 1941). Now, therefore, cultivated varieties of primulas are often uniformly of the pin-type. Such changes towards inbreeding will of course lead to sudden increases in the frequency of recessive types appearing (see Ch. VI).

This movement towards inbreeding has depended on unconscious selection equally with wind-pollinated cereals. Hermaphrodite wheat and barley have in this way become strict inbreeders. But maize with its separate male and female spikes could not become an inbreeder by any natural means. The selection which produced its large ears must have been conscious. And inbreeding also (which became the basis of Henry Wallace's hybrid maize) had to wait for the consciously and indeed elaborately planned plant breeding of this century.

(*b*) *Outbreeding replaces Inbreeding.* Strange species are brought together and permitted to cross, especially when flowering times are changed; in *Fragaria* and *Rosa, Aesculus* and *Spartina* examples are given in Section 7. Again, unfavourable climatic conditions may lead to male-sterility. This compels a normally selfed plant to cross if it is to reproduce at all, e.g. in cereals, and in tomatoes.

(*c*) *Secondary Changes.* Often in crosses between diploid species, triploid progeny alone are viable, e.g. in *Larix, Rubus, Apium* (celery), *Sonchus, Collinsia, Saccharum* and *Lolium*. Contrariwise, on self-fertilization, self-incompatible diploids will give only triploid progeny, e.g. in *Tulipa* and *Pyrus* (Lewis, 1943). Unreduced diploid gametes are favoured at the expense of the reduced haploid ones wherever an obstacle is placed in the way of the normal process of reproduction. Thus any change in the breeding habit boosts the production of triploids.

Changes which occur inevitably in cultivation thus modify or even violently reverse the character of the breeding system as it existed in the species before cultivation. This change in turn leads to the production of new variants by combination or segregation and also upsets the general genetic adaptation, that is the genetic system, of the species or group (Mather, 1943). And those conditions that favour hybridization and polyploidy often conduce in turn to more hybridization and more polyploidy. Finally, as in *Saccharum* or *Citrus*, some form of apomixis may bring the story to a close by stopping the recombination and segregation on which the whole development has depended.

In short, the method of reproduction of crop plants, whether self- or cross-pollinated, or vegetatively or asexually propagated and the extent to which these methods have been altered by acclimatization, have determined the condition of their variation. At the same time the processes of cultivation have altered, as we have seen, the conditions of selection. And it is these concealed conditions which have directed the development of cultivated plants.

The improved nutrition under cultivation has had a direct effect only on individuals in producing better growth which is not inherited. Nutrition, however, has had indirect and unintended effect of great importance. In producing this better growth and suppressing the competition of wild unwanted plants, it has made possible the survival of genetically more vigorous plants of the crop species which would have been (and still are) discouraged or eliminated under natural conditions. Cultivation has thus altered the conditions of unconscious selection in the direction required by man.

These improved conditions are also more standardized. The wild plant has to be prepared for great variations of season and soil. It is prepared we find in two ways, by the variety of individuals in the population and by the plasticity in each individual's reaction. In cultivated plants the need for both the variety and the plasticity is reduced. Plants have certainly lost their variety under domestication as we shall see later. Their loss of plasticity on the other hand is harder to show. In the tomato, however, an ingenious test has been successfully applied.

In cultivated tomatoes as in so many other plants the fertility of triploids even when crossed with the pollen of diploids is greatly reduced owing to the heavy elimination of the unbalanced embryos and seedlings. Instead of setting about 100 seeds per fruit from 100 ovules as in the diploid, the cultivated triploid sets an average of 0·5 seeds per fruit, and of these 61 per cent. germinate to give seedlings with unbalanced chromosome numbers. Rick however found that a primitive type of tomato in its triploid form had an average of 4 seeds instead of 0·5 seeds per fruit and a proportion of 88 per cent unbalanced seedlings instead of 61 per cent. It seems likely that chromosome balance has become more rigidly adjusted in the cultivated form than in its wild ancestor. Some kind of plasticity has been reduced by the processes of unconscious selection under cultivation.

Probably it is a consequence of all processes of unconscious selection under domestication applying to all plants and animals and to man himself, that individuals lose their readiness to flourish under a wide range of conditions to which they are no longer in danger of being subjected.

7. THE INVENTION OF PLANT BREEDING

Only recently, if we except his work in improving maize, has man come to undertake consciously controlled plant breeding.

During the seventeenth and eighteenth centuries a new method began to transform the still primitive world of cultivated plants. The long stabilized mixture of conscious and unconscious processes, which had in a few thousand years yielded the vast improvements that we can infer, began to give place to systematic processes of three kinds. First, there was selection, not casual or unconscious but persistent and unremitting. Secondly, there was propagation, not parochial but national and even international. Thirdly, there was hybridization based for the first time on the knowledge that pollen as well as ovules contributed something to heredity. All this was in a world suddenly enlarged by the great navigations.

In this revolution the Dutch bulb breeders, whose commercial history has been described by Krelage, led the way, beginning with their tulips and hyacinths newly arrived from the east and continuing with all kinds of ornamental plants. But it was the firm and the family of Vilmorin, established in 1727, which a little later developed the same possibilities with crop plants. They began the work, continued by seedsmen ever since then, of replacing mixed, unselected, 'land races' within which natural selection was continually operating, by standard-ized, artificially selected, and inbred 'varieties' of crop plants. Later, the knowledge of the nature of the sexual process in plants promising, as it did, advances analogous to those achieved by selective breeding in animals, inspired Louis de Vilmorin, Thomas Andrew Knight and their contemporaries to make far-reaching experiments first in the selection and then in the hybridization of crop plants.

The replacement of accident by planning can be seen equally clearly in the use of selection and of hybridization. The Silesian sweet fodder beet had been picked out already in 1745 as having in one strain as much as 6·2 per cent. of sugar in its roots. It was introduced into France in 1775. By selection year after year of the sweetest beet for breeding, Pierre Philippe de Vilmorin raised its sugar content to 16 or even 20 per cent. and produced the beet which relieved the sugar shortage of France during the Napoleonic wars. Thereby he created what was virtually a new plant, a plant which has become one of the staple crops of the world today and much the most impor-tant of the four crops successively derived by selection from *Beta maritima*.

During the same period the development of the strawberry was taking place: here hybridization preceded selection and what began as an accident was improved by intent. In 1790, plants of the South American *Fragaria chiloensis* were fertilized in Brittany with pollen of the North American *F. virginiana*. This accident took place simply because the particular imported *chiloënsis* plants happened to be male sterile. Only a few years later, Keen in England was crossing and selecting their deriva-tives deliberately in order to produce the plants from which Laxton later bred the modern strawberry of cultivation.

In establishing these sudden advances ancient methods of vegetative propagation were of great value. Plants like the potato and the strawberry of course propagate themselves vegetatively in nature. In this way a single uniform individual, a clone, may succeed in colonizing the whole earth. Many trees and shrubs, like the olive and the fig, have no doubt been artificially propagated from cuttings or suckers since the dawn of agriculture. But the method to which we owe the maintenance of a large part of our fruit trees and ornamental shrubs, namely grafting, is relatively recent. There are suggestions in the Odyssey that grafting was understood in Homeric times; but these must be misinterpretations for there is no mention of grafting in the Old Testament. Yet, as Hehn points out, it probably first established itself in the Levant. The writings of Theophrastus, Pliny, Columella and the apostle Paul illustrate its westward expansion through the Mediterranean. No doubt it was in this region and at this time that our oldest vegetative varieties of European fruits had their origin. Only later did the practice spread to enrich the cultivations of the East.

The consequence of this introduction of new kinds of vegetative propagation as a means of preserving new types is shown by chromosome studies in the way we have already noticed. In the edible apples and pears of Europe, just as in the ornamental cherries of Japan, triploid forms begin to be preserved. In these flowering cherries, with a small basic number ($n = 8$) and one or two seeds to the fruit, they are quite sterile and this is an advantage. In the apples and pears with a larger basic number ($n = 17$) and ten seeds to the fruit, they are merely of reduced fertility and this also is an advantage for the fewer fruits are also the larger and the more regularly borne fruits. In modern times therefore, unknown to the breeders and propagators, the proportion of triploid plants in cultivation has steadily increased.

In the nineteenth century the revolutionary techniques of plant breeding and vegetative propagation invaded the chief crop plants of temperate climates. The heterogeneous natural populations of primitive agriculture were replaced over a large part of the earth by homogeneous standardized varieties, often varieties closely related to one another. This change brought with it is own difficulties and dangers. Nature gives an example of what happens in such circumstances. Vegetatively and apomictically reproduced species are continually arising in nature especially by polyploidy. Only rarely, however, do they establish themselves, and their success must usually be short-lived. Why? The heterogeneous sexually reproducing population always has a chance of partially surviving the attacks of disease and of parasites; and the survivors will be selected for resistance. The homogeneous population, on the other hand, whether sexual or clonal, is liable to be wiped out entirely. Its stability and its uniformity condemn it. The rigour of artificial selection for quality or vigour precludes the possibility of natural selection for resistance to disease. Thus as soon as the perfect variety is produced, whether wheat or oats, banana or potato, it is exposed to the attacks of highly adaptable and rapidly evolving insect, fungus and virus pests attacks which it is unfitted to sustain.

The same principle applies to resistance against cold and all other extremes of the environment. But these intermittent dangers present only one aspect of the consequences of altering and specializing the method of selection of crop plants. For not only the plants themselves but the breeding system also is altered with less apparent, but no less dangerous, consequences. Thus the pre-eminence of a few selected clones of self-incompatible fruits like plums and cherries has led to the large scale multiplication of single varieties or groups of closely related and inter-incompatible varieties. This has resulted not in death but merely in sterility, for which correct inter-planting provides the cure. But inter-planting in turn requires that new and genetically appropriate varieties shall be continually supplied by the plant-breeder (Crane and Lawrence, 1952).

Thus when the natural population and natural selection are abandoned they have to be replaced, not by an isolated act of selection, by by a permanent process of plant breeding, involving acclimatization, hybridization and selection, involving indeed the utilization of the whole apparatus of genetics and chromosome study.

The effect of these successive changes on cultivated plants in the world at large has been to produce a state of confusion. The problems of different crops are inherently different. But in addition, for the various kinds of crop, plant-breeding has reached different stages of development even in the same country. We have to recognize that with many of the newer crops of tropical countries, such as rubber and palm nuts, tea and cocoa, we are still at an early primitive stage with vast possibilities of hybridization, selection and standardization; and sometimes the urgent necessity if disease is not to destroy the whole of our plantations. With others again, such as coffee, citrus fruits and cotton we are at an advanced, and even the most advanced, stage.

8. MECHANISMS OF IMPROVEMENT

Chromosome numbers can be used, as we have already seen in such a genus as *Triticum*, to classify both wild and cultivated forms in genetically valid groups or species. When Sakamura discovered that the cultivated wheats fell into three groups according to their ploidy he resolved, and finally resolved, a confusion which, as we saw earlier, had endured for 7,000 years or more. In the same way when Karpechenko found that there were three chromosome series in *Brassica*, 8, 9 and 10, the past history and future possibilities of the confused group of cabbages and turnips, swedes and kales became clear. The reason for this is that different basic numbers and different stages in the polyploid series imply inter-sterility and therefore determine successive steps in evolution. We have now reached a point at which this principle can be universally applied. We can use the study of the chromosomes in finding out how the evolutionary steps have been made. First, in the development of cultivated plants, we can sort out the several ways in which they have arisen. Of these four may now be formally distinguished:

1. By selection amongst genetic differences existing, or mutations arising, within a single mating group or genetic species:

Rubus idaeus ($2x = 14$) → summer raspberries
Pisum arvense ($2x = 14$) → *P. sativum*
Linum bienne usitatissimum ($2x = 30$) → linseed → flax
Beta maritima ($2x = 18$) $\begin{cases} → \text{spinach beet} \\ → \text{garden beet} \\ → \text{mangold} \\ → \text{sugar beet} \end{cases}$

Hordeum spontaneum ——→	*H. vulare* (Zohary 1959, 1960)
brittle hairy rachis,	non-brittle less hairy rachis
lax ears two-rowed	compact ears six-rowed
long awns	shorter awns

2. By polyploidy within such a group, giving reduced seed fertility:

(a) AUTO-TRIPLOIDY

> *Citrus aurantifolia* ($2x = 18$) → seedless Tahiti lime
> *Chrysanthemum frutescens* ($2x = 18$) → marguerite
> *Pyrus malus* ($2x = 34$) → Blenheim Orange, 1750, etc.
> *Tulipa gesneriana* ($2x = 24$) → Zomerschoon, 1600, etc.
> ~~*Colocasia antiquorum* ($2x = 28$) → Taro yams~~

(b) AUTO-TETRAPLOIDY

> *Vitis vinifera* ($2x = 38$) → Muscat gigas grape, 1900, etc.
> *Primula sinensis* ($2x = 24$) → giant forms, c. 1900
> *Pyrus communis* ($2x = 34$) → Double William, 1935, etc.
> Summer raspberries ($2x = 14$) → autumn raspberries, 1917

(c) SUCCESSIVE

> *Hyacinthus orientalis* ($2x = 16$) → 1700–1800, $3x$
> $\qquad\qquad\qquad$ → modern varieties, $2x$–$4x$
> *Populus tremula* ($2x = 38$) → gigas clones, 1936, $3x$
> $\qquad\qquad\qquad$ → seedlings, 1940, $2x$–$4x$
> *Ananas commons* ($2x = 50$) → pineapples, $4x$ → $3x$
> *Primula malacoides* ($2x = 18$) → $4x$ → $6x$
> *Chrysanthemum maximum* ($10x = 90$) → $15x$ → ca $20x$
> *Musa acuminata* ($2x = 22$) → $3x$: AAA, $4x$: AAAA
> *M.a.* × *balbisiana* ⟶ $\begin{cases} 2x\text{: AB} \\ 3x\text{: AAB, ABB} \\ 4x\text{: ABBB} \end{cases}$

3. By selection amongst segregates after crossing between two such groups:

(a) BOTH PARENTS DIPLOID

> *Ribes vulgare* × *R. petraeum* × *R. rubrum* ($2x = 16$)
> $\quad$ → red currants, before 1600
> *Petunia axillaris* × *P. integrifolia* ($2x = 14$) → garden petunia, 1834
> *Streptocarpus rexii* × *S. parviflora* × *S. dunnii* ($2x = 32$)
> $\quad$ → garden streptocarpus, 1890

(b) BOTH PARENTS POLYPLOID

> *Fragaria virginiana* × *F. chiloensis* ($8x = 56$)
> $\quad$ → garden strawberries, 1790–1850

4. By polyploidy following such crossing:

(a) TWO-SIDED DOUBLING

> *Aesculus pavia* ($4x = 40$) × *Ae. hippocastanum* ($4x = 40$) → *Ae. carnea* ($8x = 80$), c. 1800
> *Spartina stricta* ($8x = 56$) × *S. alterniflora* ($10x = 70$) → *S. townsendii* ($18x = 126$, c. 1870)

Phleum pratense ($2x = 14$) $\times$ *P. alpinum* ($4x = 28$) $\rightarrow$ American
   timothy ($6x = 42$)
Nicotiana sylvestris ($2x = 24$) $\times$ *N. tomentosa* ($2x = 24$) $\rightarrow$ *N.
   tabacum* ($4x = 48$)
Brassica oleracea ($2x = 18$) $\times$ *B. rapa* ($2x = 20$) $\rightarrow$ swede or ruta-
   baga ($4x = 38$), c. 1680 (introduced to England 1781)
Triticum dicoccum ($4x = 28$) $\times$ *Aegilops squarrosa* ($2x = 14$) $\rightarrow$ *T.
   spelta*, *T. vulgare*, etc. ($6x = 42$)
Prunus divaricara ($2x = \times 16$) *P. spinosa* ($4x = 32$)
                                    $\rightarrow$ *P. domestica* ($6x = 48$)
                                    $\rightarrow$ *P. insititia* ($6x = 48$)

(*b*) ONE-SIDED DOUBLING

Delphinium nudicaule ($2x = 16$) $\times$ *D. elatum* ($4x = 32$) $\rightarrow$ *D.
   'ruysii'* ($4x = 32$), 1929
Rubus idaeus ($2x = 14$) $\times$ *R. ursinus* ($8x = 56$) $\rightarrow$ loganberry
   ($6x = 42$), 1880
Rubus rusticanus inermis ($2x = 14$) $\times$ *R. thyrsiger* ($4x = 28$)
                    $\rightarrow$ John Innes Berry, thorned ($4x = 28$)
                    $\rightarrow$ Merton Thornless, 1935
Saccharum officinarum ($8x = 80$) $\times$ *S. spontaneum* ($8x = 64$)
                    $\rightarrow$ Coimbatore Cane, Co 205 ($12x_2 = 112$), 1916

5. Asexual propagation of a sterile hybrid clone.

Various cultivated forms of *Allium* ($2x = 16$, Vosa 1969) and of *Narcissus*
(Wylie 1952). Note: this situation is found in nature with hybrids of *Opuntia*
(V. & K. A. Grant 1971).

The simplest of these changes are, of course, those giving
auto-polyploids, but even they provide valuable lessons. The
triploids, as we saw, are found only where seed production is
of little importance or needs to be depressed.

Auto-tetraploids show similar properties; but they also reveal
that selective character which we inferred in auto-tetraploid
wild plants. Lewis has shown that garden raspberries exist in
two chromosome forms, diploid and tetraploid and also in
two gene-forms, summer-fruiting and autumn-fruiting. The
diploids are mostly summer-fruiting, the tetraploids entirely
autumn-fruiting. The reason for this is that tetraploidy ruins
the pollen of the summer flowers but not of the autumn
flowers. Possibly temperature and other conditions reduce
chiasma formation and create functional diploidy in the tetraploid raspberry just as a chromosome change does in the tetraploid wheat. In any case tetraploid raspberries have been unconsciously selected among the progeny of autumn-fruiting

diploids. This is an example of the principle of mutual selection which we saw governing the change in vegetative propensities of polyploid species.

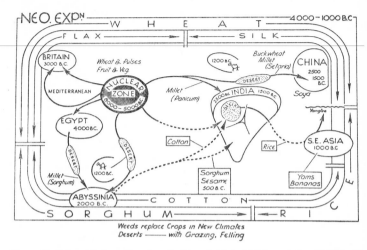

FIG. 46. Map-diagram showing the Bronze Age division of the Old World by Regions in respect of crops following the growth of deserts and the domestication of the camel (cf. Fig. 42).

The classification of the allo-polyploids is, of course, misleading unless it is understood to be diagrammatic. The more recent cases are precise and authentic, the more remote, approximate and hypothetical. The distinction between the *Ribes* and *Beta* types of origin is, from what we saw earlier, arbitrary. The *Triticum, Prunus, Brassica* and *Nicotiana* cases are attempted reconstructions of an earlier origin. But in one principle they are equally correct: they show the different processes by which variability has been released and exposed to selection. With the species crosses which are here summarized this release occurs in two ways subject to the number and relationship of the chromosome sets concerned.

1. *Primary Segregation:* from pairing of chromosomes of opposite parents, e.g. in a diploid cross between two diploids.

2. *Secondary Segregation:* from pairing of chromosomes

178

within the complements of one or both polyploid parents in their offspring.

Thus, where no doubling has taken place, two tetraploid species with sets AABB × CCDD will give a hybrid ABCD which will produce gametes either by A—C and B—D pairing (primary segregation) or by A—B and C—D pairing (secondary segregation). The relative importance of the two types of segregation is bound to differ in different tetraploid crosses. But under experiment no one has yet attempted to assess it.

Where doubling of the chromosome number has taken place on both sides or after crossing, segregation of both kinds is more or less suppressed; every chromosome has an identical partner, and, so far as it regularly takes this partner, the new form is fertile and true-breeding. Perfect stability is never attained, however, in practice; new polyploids and new crosses between polyploids always release some unexpected variability —unexpected if we have not studied their chromosome be- haviour at meiosis. No doubt also in many new high polyploids like *Chrysanthemum maximum*, *Spartina townsendii* and *Sac- charum officinarum* some of the variation that arises is through whole chromosomes being lost in irregular meiosis and giving unbalanced progeny which survive.

With one-sided doubling an intermediate segregation should take place. Bread wheat types have been produced as above by two-sided doubling; but doubling on one side might be even more effective. Thus we might suppose that if the cross *Triticum dicoccum* (AABB) × *Aegilops ovata* (CCDD) had doubled on the *Triticum* side it would give a hybrid AABBCD. This combination although not attempted in experiment may have happened in nature.

A description in terms of formal species names and a single process is likewise misleading as applied to the origin of the sugar cane. Here we are concerned with a complex of wild forms of *Saccharum* with chromosome numbers between 48 and 112, classified under the names *spontaneum* and *robustum*, and distributed between Turkestan and Polynesia. By selection for size and sugar content, and of course increase of chromo- some number, these forms have given rise to the cultivated species called *barberi* in India, *sinensis* in China, and *officinarum* in Java and the Pacific. In recent years these have all been crossed together and with the wild forms and, with further increase in chromosome number up to 148, have finally given us the prodigies of Passeroën and Coimbatore.

In these descriptions we have therefore to remember the genetic diversity that may be covered by a species name that has been given usually without genetic knowledge. We have to know what kind of mating method, chromosome organizaion and internal discontinuity may be concealed by that name. In these circumstances we are bound to relate the kind of origin or development (single or multiple, sudden, gradual or successive) and the kind of variability (high or low, continuous or discontinuous) with the hybridization and polyploidy which chromosome studies can reveal. We may well connect the invariability of *Morus nigra* as compared with *Morus alba* not, as De Candolle did, with its newness in cultivation but rather with its high allo-polyploidy. On the other hand, the variability of *Prunus laurocerasus* we may well connect with its high auto-polyploidy rather than with any antiquity in cultivation.

9. OIL AND FIBRE PLANTS

We have now considered a number of staple crops in relation to the principles which seem to have governed their domestication. These principles concern their geographical origins and movements, their modes of propagation and their breeding systems, their chromosome changes and the sexual fertility connected with them, their selection by conscious and unconscious processes in the course of commerce and cultivation. Let us now see how these principles apply to particular plants. Let us choose moreover a group classified by its uses but characterized by divergence in their development so as to throw light on the most complicated situations.

Many plants have been selected for diversified uses within a limited range. Such are the beet and cabbage families of vegetables. Each of these has, however, continued to fall within one class as either a root crop or a green crop. It is in the divergence of oil and fibre crops from the same common origins that we get the most instructive situations in the human control of the evolution of crop plants. We may pick out three for special treatment: hemp because we can see the evidence of its origins in the cultural details today; cotton because its movements and classification are more complex and better known than those in any other plant; linseed because its early history is the best verified archaeologically.

(i) Hemp

Wild hemp, the dioecious species *Cannabis sativa*, spreads today over a region from the lower Volga and the Caucasus in the west to the Altai mountains in the east. It is cultivated in central Asia for its oil seeds, in India for the drug extracted from its female flowers, leaves and bark, and in southern Europe (but no longer in Eastern U.S.A.!) for its fibre. The first historical record of this expansion we find in Herodotus who refers (4·74) to the introduction of hemp to the Greeks by the Scythians in the fifth century B.C.

The importance of hemp for us is that Vavilov (in 1926) used it as an example of how we can distinguish between a genuine wild plant and an escape from cultivation. At the same time he pointed out the selective processes at work in cultivation. The wild hemp, he found, requires soil as rich as the cultivated varieties, liking the winter camps of the nomad tribes in the Volga and Altai regions, as much as the valley bottoms of the Caucasus. Seeds, he noticed, were often still collected in the wild for sowing in the garden. The evidence that they were genuinely wild hemp was of four kinds:

(1) germination of seed is slow and irregular,
(2) the perianth persists as an outer husk round the fruit developing a disguising pattern,
(3) the fruit has oil glands outside the pericarp; these attract various insects which remove and distribute it,
(4) the fruit shatters and distributes the seed.

These characters are of the kind familiar to us as present in wild grasses and lost in cultivated cereals. But here, as

Vavilov points out, the wild plant is following man around in his habitations and 'proffering its services'. Although, therefore, hemp has been cultivated for the production of seed for 3,000 years it is still being brought into cultivation from a wild state. This reminds us of the importance of continuous and migrating domestication, spread over a vast range of time and space, as opposed to the sudden and local introduction which Vavilov himself assumed as a general principle.

Pursuing the problem further, Vavilov noted that in the oasis of Khiva stems of hemp were still rubbed out dry. This was the original method of preparation. Only later was the wet rotting or retting, so well known in flax, used for extracting the fibres. Later still, no doubt, when burning the stems, men noticed the narcotic effect of the smoke. In southern races and perhaps also inherently in warmer climates the narcotic substances increase and offer the incentive for the preparation of hashish. Last of all came the extraction of seed oil from the seed and the world-wide cultivation of hemp.

(ii) Cotton

The history of cotton has come to be understood owing to the discoveries of modern archaeology and the prolonged investigations of plant breeders, assisted by its contrasted chromosome constitutions. It tells us something not only in all these directions but also in regard to one of the fundamental questions of chromosome botany, that of the relations of changes in plant habit and in geographical distribution. We owe to Sir Joseph Hutchinson the fullest account which we may summarize here.

Ancestral species of linted cotton must have existed in Africa and India as well as in Mexico and Peru. But today these ancestors are mostly lost and wild cotton with linted seed survives only in the bush veld of southern Africa as *Gossypium herbaceum africanum*. It must, however, have come into cultivation in southern Arabia (where it still survives escaped from ancient cultivation). Here it was probably found by the earliest Sumerian traders on their way to the Indus valley ports. Textiles made from this coarse early cotton have been found in Mohenjo-daru and dates about 2000 B.C.

Cotton was not, however, in this period introduced to the cooler western lands. It disappears from sight for nearly a thousand years. During this time it seems likely that from the Indus it spread east into the Ganges valley. There, perhaps by crossing with native wild forms which have been lost and by later selection, it developed into the species now known as *arboreum*. Later with the spread of cultivating people it passed eastwards into Indonesia and at the same time returned overseas westwards back to Africa. There at Meroë in 400 B.C. it was the first cotton fibre to be spun and woven in Africa.

All this history, we must recall, was the history of a slow-growing perennial woody plant too sensitive to frost to be cultivated far outside the tropics, for example, in Sumeria or Egypt. So far it had no doubt been slowly selected, unconsciously for self-fertility, more consciously for the quantity of its fibre. Now, however, during the first millennium B.C. a new, and this time perhaps conscious selective change began to make itself felt in Indian cottons. Cultivators began to look for a quicker yield by early maturity. In doing this they produced strains with an annual habit. This physiological upheaval led to a geographical revolution. Cotton for the first time could move right out of the tropics. Its cultivation was soon extending into the colder regions of Persia and Turkestan. There improved annual forms must have developed: slowly perhaps for the consequences were seen only long afterwards. It was only in modern times, in the eighteenth century, that annual forms of Persian cotton were brought back into India, indeed once more into the Indus Valley, where cotton thereby was transformed from a small domestic garden crop to a staple field crop.

Meanwhile in the seventh century, the Indian cotton in Indonesia was undergoing a parallel transformation. It was carried to south China where it was grown as an ornamental perennial. Here again rapid maturing types seem to have been selected in the eleventh century and grown for fibre and their annual habit enabled them to colonize the colder north of China.

This completes the story of the much-migrating Old World Cottons. They are all connected with or derived from the African cotton. They are all diploids and all inter-fertile races of one diploid species. The New World Cottons, on the other hand, are all derived from what were two stocks when man arrived. Both of these were tetraploid. One was Mexican (*G. hirsutum*), the other Peruvian, (*G. barbadense*). Both of them were again perennial woody and non-hardy tropical species. The cultivation of these two species began probably independently, about 3000 B.C. Both of them by the time of the discovery had spread to the northern seaboard of S. America and the West Indies where they overlapped in distribution.

It is worth noting that, since both the ancient American civilizations began in the tropics, cotton came into them early as flax did in the Old World. In the earliest remains in Peru at Huaco Prèta cotton is found with beans, cucurbits and ground nuts (from N.W. Argentina): it precedes pottery and maize.

Mexican cotton developed annual types but these did not colonize the southern U.S.A. until modern times in the eighteenth century. And only in 1865, after the civil war, did they begin to supplant Old World cottons in Central Asia and Africa. At the same time by way of the Philippines they penetrated Cambodia and India and soon began to dominate Indian cultivation.

Peruvian cotton began its expansion last of all, but, on account of the fineness of its fibre has travelled as fast in the last two centuries as the Mexican. Rapid maturing forms were taken from the West Indies to the Sea Islands of South Carolina in 1786. There good annual types were selected with the name of Sea Island Cotton. Meanwhile a perennial Peruvian cotton passed into Nigeria and by the slave route to Egypt. Here its cultivation after 1820 was so successful that it led to a change of the system of Nile irrigation which had sufficed for nearly 5,000 years. The seasonal flooding system was redesigned to provide permanently for whole-year perennial cultivation. Thus Egyptian agriculture was incidentally revolutionized. Annual Sea Island cotton was however soon introduced from America. It met and crossed with its relative from Nigeria and produced with selection the modern cottons which are today the staple of the Egyptian industry.

The whole history of cotton, its biological evolution on the one hand, and the agricultural, industrial and social effects of that evolution on the other, has been bound up for 5,000 years with the use of its fibre for the making of fabrics. It was not until 1783 that, on the analogy of linseed and other oil seeds, the extraction of the seed for oil and cattle food was first suggested (and a medal offered for the invention of a process by the Royal Society of Arts in London). And only sixty years later was the idea put into practice in France. Some 10 million tons of seed are now extracted every year but in the use of the cotton crop the oil is secondary to the fibre.

(iii) Flax and Linseed

As a contrast to the history of cotton the conclusions reached by Helbaek on the parallel history of flax and its seed crop lint-seed or linseed are illuminating. The probable wild ancestor survives as a single variable species, *Linum bienne*, whose forms range from a coastland Mediterranean-Atlantic race, which is perennial, to a race around the borders of Mesopotamia which is a biennial or a winter annual in habit. It was from these forms that winter-annual oil-seed producing forms were first cultivated and selected in Kurdistan around 5000 B.C. Later in Egypt forms were selected as summer annuals which are the origin of the modern fibre flaxes. It is these, as with cotton, which have allowed the development of flax growing in temperate climates. Linseed cultivation, on the other hand, for the extraction of the oil has stayed in the warmer climates where it began.

In their selection for rapid maturity and annual habit *Linum* and *Gossypium* are thus closely similar. But in the development of their uses and their relation to climate they are quite contrasted.

Cultivated plants, in general, have given us a clearer view of the principles at work in the evolution of wild plants. But there remain certain genetic problems on which their evidence is too remote to be certain. In order to settle these questions we must closely examine the best understood of all plants, those most recently introduced into cultivation, the ornamental plants.

VI. ORNAMENTAL PLANTS

OVER 9,000 years have passed since the successful cultivation of crop plants led to the first settlements appearing on the hillsides of the Fertile Crescent. But scarcely for the last 4,000 years has man's superfluous prosperity arising by this means encouraged him to go further and cultivate plants merely for decoration. The culture of flowers arose independently in the Old World and in pre-columbian America. Its early pre-eminence and variety in India and China as compared with the Mediterranean was favoured by many conditions, by the artistic sensibilities of the eastern peoples, by the warmer climates of their countries and also, no doubt, by the more brilliant colours of the native flowers. To secure the services of the available Lepidoptera for pollination, as Beale and others have pointed out, tropical plants have needed to develop all the brilliant resources of the pelargonidin series of anthocyanins.

Religion has often seemed to encourage this development. But if the olive was sacred to Pallas Athene and its leaves were used to crown the brows of Olympic victors, it was because the tree was already highly regarded for its wood and shade and fruit and oil. And if Buddhism required the people to offer flowers rather than to sacrifice blood it was probably because the people in the countries where Buddhism spread preferred it so. Certainly, we must admit that Buddhism organized a cult of flowers which is still most highly developed in those countries where this religion has flourished. On the other hand, throughout the Middle Ages Mohammedan power barred the way between Europe and the flower-cultures of the East. It was only in the sixteenth century that decorative plants began to make their way westwards to Italy, Holland, England and France.

In swift but significant succession the university gardens in these different countries, Padua (1545), Leyden (1587), Oxford (1621) and Paris (1635) opened to receive the new botanic wealth which, ever swelling as new lands were explored, transformed them from medieval Physick Gardens with a few hundred species of 'simples', first into Exotic Gardens, and then into Botanic Gardens. During the reign of George III and the rule of Sir Joseph Banks, there were brought into England 7,000 new plant species, far more indeed than all previous generations of men had known. Inevitably most of these species were ornamental.[1] Most were brought in contact with their hybridizable relatives for the first time. It is no accident, therefore, that the second curator of the Oxford Botanic Garden, Jacob Bobart the Younger, undertook experiments in crossing male and female plants of dioecious *Lychnis*. Nor that in the Tübingen Botanic Garden a few years later Camerarius solved the problem of their sexual reproduction with monoecious maize. Nor that in the same place Koelreuter made the first series of plant hybrids by crossing ornamental plants, species of *Nicotiana* and *Dianthus*. From experiments on crossing such species there arose both the theory of genetics and the practical development of ornamental plants.

Thus, while the improvement of our staple crop plants is remote and its method sometimes conjectural, the improvement of our ornamental plants is recent and well-recorded. It is to this fact, and to the immense number of species concerned, that the ornamental plants owe their botanical and genetical interest.

A number of conclusions we have already derived from their history. It is now worth our while to consider a few examples in greater detail as illustrating general principles. These may be grouped under four heads according to their diverse and indeed contrasted modes of origin:

A. From a single diploid parent

    1. *Lathyrus odoratus*
       (Crane and Lawrence, 1952)
    2. *Primula sinensis*
       (de Winton, 1929;
       Mather and de Winton, 1941)
    3. *Hyacinthus orientalis*
       (Darlington *et al.*, 1950)

[1] In the same year, 1789, the two most important of all ornamental species arrived from China in the Port of London: *Rosa chinensis* and *Chrysanthemum indicum*.

B. From several diploid parents	4. *Narcissus*
	(Fernandes, 1951; Wylie, 1952)
C. From diploid and polyploid parents	5. *Iris*
	(Stern, 1946)
	6. *Rosa*
	(Täckholm, 1922; Hurst, 1941; Wylie, 1954, 1955; cf. Willmott, 1910–14)
D. From a complex of polyploids	7. *Chrysanthemum*
	(Dowrick, 1953)
	8. *Dahlia*
	(Lawrence, 1931)

1. *Lathyrus odoratus*

Seeds of the sweet pea were sent to Enfield near London from its native country of Sicily in 1699. There have never been any derivatives from crossing this species with any other. It has recently been crossed with *L. hirsutus* but the progeny have had no connection with the garden varieties. Indeed, in England the natural crossing of the sweet pea was entirely replaced by selfing. During the period up to 1886, three or four Mendelian recessive mutations appeared. In the next twenty-five years a cataract of changes followed from whose Mendelian recombination the tall large-flowered, somewhat less scented, sweet peas of modern horticulture were selected.

Tetraploid seedlings have been found. Owing to the high frequency of interstitial chiasmata quadrivalents are regularly formed but are clumsy in movement and irregular in segregation. The plants are therefore highly infertile and useless in breeding. All variation is thus within the diploid stock and, so far as is known, without any visible change in the chromosomes. How do we account for the long-delayed break in the stability of this species? It seems that by selecting mutants watchful breeders ultimately succeeded in selecting stocks more mutable than the wild species. Only when the mutants had appeared was crossing undertaken, the object being to recombine the changes to give desirable forms.

2. *Primula sinensis*

This plant was brought to London twice, in 1820 and 1826, from gardens in Canton. No wild or cultivated relative has ever been found in China and there is no other species of *Primula* with which it is known to cross.

The earliest seedlings varied in two respects. They were heterostyle and depended on inter-crossing of the pin and thrum types. They were also of both the frilled and one-tier or *sinensis*, and the knicked and two-tier or *stellata*, forms. Each of these variations, although no doubt supergenic in origin, behaves as a Mendelian difference. Three years, that is two generations, later other recessive Mendelian differences affecting flower colour and leaf shape began to appear. As time went on these variations became more and more frequent so that to day about a million recognizably different types could be grown all with apparently the same chromosomes.

These new forms, it is again obvious, arose by spontaneous mutation and not as the result of crossing. Indeed the practice in cultivation once more was to allow crossing to be replaced by the regular selfing of pin plants. This change was inevitable for a plant which had been pollinated by insects in the open in China and was now flowering in a glass house in winter in England. But it has introduced a new type of selection which, as we saw, has led to increasing self-fertility of the pin type.

In addition tetraploids arose and were recognized as such early in the 1900's. Being auto-tetraploids they are less fertile than the diploids. Their chiasmata being entirely terminal the segregation is less irregular and the infertility is less extreme than in *Lathyrus*. Their seed is therefore merely somewhat more expensive than that of diploids. They cross with the diploids with some difficulty and only as female parents. Most of the progeny are sterile triploids but a few, through the action of unreduced pollen, are tetraploids. Such variation of diploids as was wanted in the tetraploids has by this means been transferred to them.

In the course of a little more than a century, a wide range of variation in external form and in chromosome complement has thus arisen within a uniform stock in which crossing has been largely suppressed. Within *P. sinensis* we now have two nearly intersterile forms, diploid and tetraploid, potential new species. Within this group of manifestly common descent we can now have a certain kind of hybridization and thereby produce manifest sterility. But the diversity that has arisen has done so spontaneously. It is the cause and not the consequence of any hybridization that may occur.

We must note that unbalanced individuals (trisomics and so forth) which do not occur at all in *Lathyrus odoratus* are rare and probably less vigorous in *Primula sinensis*. In neither species do they make any contribution to the origin of new varieties.

3. *Hyacinthus orientalis*

The hyacinth was introduced to Holland from the Lebanon or Taurus mountains about 1560. It still exists with a very uniform character in its native habitat. It has the same loose spikes of blue flowers that were illustrated in the herbals in the seventeenth century. And it has the same diploid complement of 16 chromosomes that is still revealed by the older varieties in cultivation. Moreover, the other species of *Hyacinthus* have entirely different chromosome complements and there is no other species with which any garden hyacinth has crossed. This is one of those rare cases where nature conforms diagrammatically with the accepted laws of systematics.

The hyacinth, like the sweet pea, and *Primula sinensis*, thus began from uniformity and, without hybridization, developed of itself, if not an immense at least a considerable diversity. By 1768 there were nearly 2,000 named varieties and the next hundred years saw as many more. How did it happen?

In the first century white, pink and double forms appeared probably as seedlings; probably, again, as a result of the unaccustomed self-fertilization of solitary plants in Dutch gardens. Thereafter new variations came thick and fast. Lilac and then yellow were added to the range of colours. The shape of the flower spike changed; the numbers of flowers and leaves increased. Without knowing it the breeders had selected triploids ($3x = 24$). Now, as we saw earlier, the haploid set in the hyacinth, unlike that in *Lathyrus* or *Primula* is not made up of highly dissimilar chromosomes balanced against one another. Each of the 8 chromosomes (as Mather has put it) is similarly and separately balanced in itself. And 5 out of 8 chromosome types can be separately counted (Fig. 47).

The triploid hyacinths, unlike most triploids among flowering plants, are, for these reasons, sexually fertile. They gave the whole range of chromosome numbers from 16 to 32 in their progeny. The different numbers not merely lived but many of them made successful garden varieties. At first the Dutch breeders selected the lower numbers but later, it seems, by gene recombinations, the plant became adapted to giving vigorous results with the higher numbers. In the present century varieties with 31 and finally 32 chromosomes began to appear and to be highly recommended in the breeders' catalogues (Fig. 48).

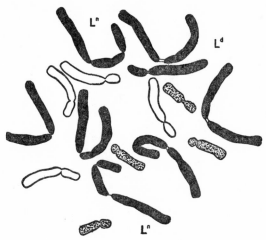

FIG. 47. The chromosome complement of a hyacinth 'Madame du Barry' with a deficient Ld chromosome which arose as a bud sport from a normal diploid: $2n = 6L + 2L^n$ (black) $+ 4M$ (white) $+ 2S_1 + 2S_2$ (stippled) $= 16$ (after Darlington, Hair and Hurcombe 1951).

Today, owing to a change of fashion in the last thirty years, only about a hundred hyacinth varieties survive. Of these only two possess single chromosomes which are slightly changed from the types recognizable in the wild species. All the rest

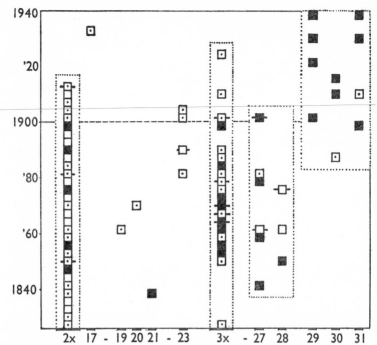

FIG. 48. Diagram showing the dates of origin of hyacinth varieties with different chromosome numbers (2x, 17 . . . 31). Black squares are blue varieties; dotted, red; and plain, other colours (after Darlington *et al.* 1951). Since 1951 one variety, Blue Giant, has now been identified with the tetraploid number 32 (Fogwill, unpub.).

have the standard complement merely varying in numbers. In the hyacinth, therefore, diversification comparable with that needed for the origin of a new species had been accomplished by a change towards inbreeding accompanied by a change in the goal of selection. Once variation had appeared selection was able to make use of a combination of gene mutations with changes in the numbers of whole chromosomes. But there were no serious changes of chromosome parts, no changes of gene arrangement or proportion such as would be indicated by inversions or interchanges. And the deficiencies and duplications of segments which may be responsible for some of the mutations, are too small to be seen.

4. Narcissus

This genus contains some twenty species distributed around the Mediterranean. There are two groups with two basic numbers: *N. tazetta* and its relatives with $x = 10$ (or occasionally 11), and the bulk of the genus with $x = 7$. One species may conveniently be separated from this main body: *N. bulbocodium* is alone in having chromosomes small enough to allow of a polyploid series in nature ($2x$, $3x$, $4x$, $5x$, $6x$). Having a *Hyacinthus*-like disregard of balance it also fills the whole range from $2x$ to $3x$, and has up to 4 heterochromatic B chromosomes as well. But *N. bulbocodium* has played no part in the origin of the garden forms.

Narcissi have been cultivated both as ornamentals and for medicinal purposes since classical times. Until rather less than a hundred years ago, the forms commonly grown consisted of the following kinds of plants:

   (i) The wild species, many of them quite variable.

   (ii) Mutants of these, perhaps found wild, such as double-flowered forms.

   (iii) Polyploid individuals occasionally selected no doubt for their increased size, e.g. *N. pseudo-narcissus hispanicus*, a sterile triploid, no doubt arising sporadically in nature, but propagated vegetatively only in cultivation.

   (iv) First generation products of crossing between species. These had occurred sometimes in the wild, e.g. *N. bernardi*, *N. intermedius*, but more commonly when cultivation had brought the species into contact with each other.

The origin of the natural hybrids, long described as 'species', was demonstrated by Dean Herbert when he made the first controlled crosses of *Narcissus* species in 1843. Herbert's conclusions have been confirmed and greatly extended by studies of the chromosome complements of the parents and the putative hybrids (Table 22). Most of these hybrids have proved to be sterile, with the notable exception of *N. pseudo-narcissus* by *N. poeticus*. It was from this cross that the modern breeding began.

Following Herbert's demonstration of the hybrid origin of many commonly grown species deliberate hybridization began chiefly in England and Holland. The four principal groups to be developed were: (1) Trumpet daffodils, which arose purely from the diploid *N. pseudo-narcissus* of Western Europe; (2) Poets' narcissi which arose from the diploid *N. poeticus*, with a more easterly distribution; (3) large-cupped and (4) small-cupped groups from hybrids between them.

TABLE 22

THE ORIGIN OF HYBRID SPECIES AND SPECIES HYBRIDS IN NARCISSUS: STERILE(S) FERTILE(F) (after Fernandes 1951, Wylie 1952)

Parents with gametic nos.		Hybrids with zygotic nos.	S	F	Wild or garden
pseudo-narcissus (7)	× *poeticus* (7)	*bernardi* (14)		2x	W—
pseudo-narcissus (7)	× *triandrus* (7)	*johnstoni* (21)	3x	—	W—
pseudo-narcissus (7)	× *jonquilla* (7)	*odorus* (14)	2x	—	—G
gaditanus (7)	× *jonquilla* (7)	*jonquilloides* (21)	3x	—	W—
poeticus (7)	× *jonquilla* (7)	*gracilis* (14)	2x	—	—G
poeticus (7)	× *tazetta* (10)	*biflorus* (17, 24)	2x3x	—	W—
jonquilla (7)	× *tazetta* (10)	*intermedius* (17)	2x	—	W—
juncifolius (7)	× *tazetta* (11)	*dubius* (50)	—	6x	W—

In the Trumpet daffodils, triploids first arose in England in the 1860's, to be followed thirty years later, and almost simultaneously in England and Holland, by the tetraploids. Higher polyploids have been found only in seedlings and since they are no larger they have not been selected. They seem to have exceeded what we may call the *optimum polyploidy* for the large-chromosome species of *Narcissus*.

Unbalanced individuals may make up to 50 per cent. or more of the unselected progeny of an auto-tetraploid *N. pseudo-narcissus*, but varieties only occasionally have numbers other than 28. Breeders have selected rigorously, although of course unconsciously, for the balanced tetraploid number. The internal balance of each chromosome cannot, it seems, be so good as in *Hyacinthus*, or in *N. bulbocodium*.

Diploid hybrids between *N. pseudo-narcissus* and *N. poeticus*

—the foundation of the large- and small-cupped daffodils—
were slightly fertile and produced a handful of diploid garden
forms. But in these groups, also in the 1860's, triploids and
tetraploids began to appear. In the tetraploids (which are
moderately allo-tetraploid) fertility was restored. New combinations of colour and form appeared and were raised to a
larger scale of size. Thus the colour of the tiny red corona rim
was transferred from *N. poeticus* to the multitude of red and
orange large-cupped and even trumpet varieties. Later, white
varieties of *N. pseudo-narcissus* were introduced into the complex and it seems to have been by interaction between their
genes and those of *N. poeticus* that pink coronas have more
recently arisen.

Four other species have been crossed with the *pseudo-
narcissus* and *poeticus* derivatives and are now beginning to play
a part in the evolution of the garden narcissi. Crosses of
cyclamineus with *pseudo-narcissus* are partially fertile, not only
in the diploids but even in the triploids. Crosses with *N. tazetta*,
N. jonquilla and the heterostyle species *N. triandrus*, on the
other hand, have so far been quite sterile and their breeding
has not progressed beyond the first generation.

The garden Narcissi show what doubling the chromosome
number can do to a hybrid when the result is intermediate
between auto- and allo-polyploidy, and when high seed fertility
is not the primary object of selection: fertility is partly maintained while segregation and recombination are only partly
prevented. The result has been to produce in a hundred years
over 10,000 named garden varieties.

5. *Bearded Iris*

All the garden bearded iris (Pogoniris group) in cultivation
before 1895 were diploid ($2x = 24$). They were derived from
species in the Trentino and Dalmatia. Crosses between these
and near-eastern tetraploid species began to be introduced in
1897. The first from Vilmorin was the expected triploid. But
soon tetraploids became more numerous and since they were
giant and fertile, and indeed the optimum polyploids, it was
from them that all modern bearded irises are derived. These

tetraploids owed their origin to occasional unreduced eggs or pollen produced by their diploid parents. The result would be tetraploid crosses half auto- and half allo-ploid on the same lines as we assumed in the discussion of hexaploid bread wheat. Thus the cross AA × BBCC would give tetraploid seedlings of the make-up AABC and by segregation of BC, which we call secondary segregation, would give the variety of forms which actually appeared in the derivatives.

The exact multiplication of the basic number is not obligatory in *Iris*: evidently differentiation and balance are intermediate in character between those of *Hyacinthus* and *Narcissus*. The ploidy is therefore approximate. In a collection of *Iris* varieties one can pick out the (approximately) diploid, triploid tetraploid and finally, pentaploid varieties by their successively larger size (Table 23).

TABLE 23

DEVELOPMENT OF THE BEARDED *Iris* (after Stern, 1946, from determinations by Randolph)

$2n$	1895–99	1900–14	1195–24	1925–34	1935–39	1940–43
$2x (+ 1)$	4	28	30	53	5	—
$3x (\pm)$	1	2	10	9	1	—
$4x (\pm)$	—	12	22	91	94	28
$5x (\pm)$	—	—	2	1	—	—

Note: before 1895 all were diploid

6. *Rosa*

The 178 species of the genus included in the *Chromosome Atlas* are distributed throughout the North Temperate Zone. They provide one of the classical examples of a polyploid series as shown by the work of Täckholm in 1922. The basic number of 7 is common to them all. Departures from strict balance are rare. They reach their limit of multiplication in nature with an octoploid, the circumpolar *R. acicularis*.

The genus contains one genetically aberrant group: the dog roses which were recognized as the Caninae section in 1863. They are native throughout Europe and Western Asia. They have been allowed eighteen species but the allowance is quite arbitrary and in fact the whole group in its whole area of distribution is in a state of flux. It includes innumerable forms which inter-breed in nature in some areas but in others are isolated and therefore largely self-fertilized. The only substantial basis of division is by chromosome number into three classes, $4x$, $5x$ and $6x$.

The reason for these conditions is chiefly the versatile system of reproduction. The basic system is subsexual but of a unique kind: seven pairs of chromosomes are formed at meiosis together with a residue of 14, 21 or 28 univalents. The univalents are lost in pollen formation, but carried unreduced by the eggs so that the self-fertilized plant breeds approximately true and, on crossing, its genes are recombined only in a small part of its chromosomes, the seven pairs. Thus, the chromosome complement may be represented as AA (7^{II}) BCD (21^{I}) in a pentaploid Canina rose. The A group is sexual and biparental. The BCD group is exclusively maternal in inheritance and vegetative or clonal in character. And the two groups, the two parts of heredity, are strictly separated in evolution.

Subsidiary to this subsexual system is a purely vegetative apomixis which probably competes with varying success in different individuals and species with the results of selfing and crossing. These peculiarities of the Caninae group have not yet been introduced into the main classes of garden roses which all seem to be normally sexual.

Garden Roses

Since Minoan times in Europe, and for nearly as long in China, roses have probably been grown in cultivation. Double-flowered forms have been found in the wild in over twenty species in Europe and China and recently in America. These

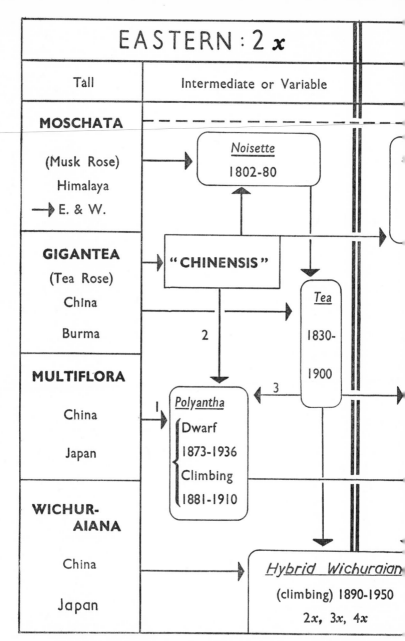

TABLE 24. Simplified diagram, derived from the work of Hurst and Wylie, of three species not known in the wild. Breeder's groups in italics underline (*Polyantha* and *Pernetiana*). The boundaries of the chromosome groups are n

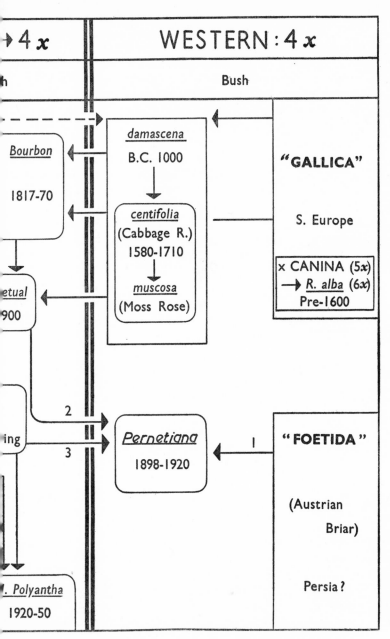

$\rightarrow 4x$	WESTERN: $4x$	
h	Bush	

Bourbon

1817-70

damascena

B.C. 1000

"GALLICA"

S. Europe

centifolia
(Cabbage R.)
1580-1710

muscosa
(Moss Rose)

x CANINA ($5x$)
→ *R. alba* ($6x$)
Pre-1600

etual

900

2

ing

3

Pernetiana

1898-1920

1

"FOETIDA"

(Austrian

Briar)

Persia?

. Polyantha

1920-50

he main groups of garden roses. Specific names in capitals. Quotation marks:
mark the stage at which a third parent was introduced into triple crosses
le lines. *Note:* the Bourbons and Hybrid Chinas are parallel developments.

especially have been selected and bred for decoration and, in the case of the damask roses, for perfume as well. It was only after the year 1802 that systematic crossing began to yield the great improvements from which modern roses have been derived. These have all come from seven Old World species, or cultivated groups to which, by an easy-going tradition, we give the names of species. The history of the crossing and improvement of the resulting garden roses is marvellously preserved for us, notably by botanical illustrations (which confirm our names), but also very often by the preservation of the critical plants in living collections. These have revealed that hybridization and selection have been carried out on a basis (unknown at the time) of several degrees of polyploidy. From the determination of the chromosome numbers in the chief groups it has therefore been possible to understand, to correct, and to explain the sequence of history.

The seven important species fall into two groups which are contrasted in their polyploidy, their geography and partly also in their habit. There are the diploids from the east which are all tall bushes or climbers. And there are the tetraploids from the west which are all smaller bush roses.

Three of these species, all eastern diploids, *R. gigantea*, *R. multiflora* and *R. wichuraiana*, present no great difficulty. They occur in the wild as white-flowered single forms. Improved, that is, double and coloured forms of *multiflora* had appeared in China while similar forms of *gigantea* had contributed to *chinensis*. These species, however, were only introduced into the history of crossing after 1800. The other species need to be separately considered:

(1) *R. chinensis* (2*x*). This is the old rose of cultivation in China no longer connected with any single wild species. It is, unlike the three wild Chinese species, a bush rose with several colours of flowers. In cultivation, perhaps after hybridization, it has been selected for double flowers of varied colours and strong scent (hence '*R. odorata*') and for a perpetual flowering habit (hence '*R. semperflorens*'). Its earliest record in Europe according to Hurst is in a painting by the Florentine Bronzino dated 1529. Being less hardy than the European roses, it did

not appear in England until it came direct from China in 1789.

(2) *R. moschata* (2*x*). The Musk Rose grows wild in the Himalayas but already in prehistoric times its double forms were cultivated and had spread east to China. It had also spread west to Greece by the time of Theophrastus. It reached England in the reign of Henry VIII.

(3) *R. gallica* (4*x*). The 'centifolia' or hundred-petalled rose of classical times and the French Rose of the Middle Ages grows wild in single forms in southern Europe and the Levant. It lends its name to the deep red roses of Lancaster and also of Tuscany. Probably by hybridization in the Mediterranean countries with *R. moschata* (which would double its chromosome number to give fertile tetraploid hybrids), this group gave the Damask Roses already in ancient times. They penetrated into England along with the Musk Roses. But the two, which had crossed in the east much earlier, in the west were now quite independent: perhaps conditions of pollination had changed. From the Damasks in turn the Cabbage Roses, the modern *R. centifolia*, were developed to the number of 2,000 varieties by Dutch breeders in the seventeenth century (1580–1710).

Modern roses with their great variety of colour, shape, habit and hardiness began to appear when the barrier between the first of these species, *chinensis*, and the other two, *moschata* and *gallica* was broken down by crossing, partly deliberate crossing. This happened in a series of steps as follows (cf. Table 24):

(i) *R. chinensis* (2*x*) × *R. moschata* (2*x*). The NOISETTE roses of tall bush habit were produced as the F_2 from this cross at Charleston, U.S.A., in 1802.

(ii) *R. chinensis* (2*x*) × *R. damascena* (4*x*). The BOURBON roses of the expected bush habit were produced as a spontaneous 3*x* F_1 whose occasional 2*x* gametes gave 4*x* plants in a back-cross to the *R. gallica* group. They were raised on the Ile de Bourbon in 1817.

(iii) From the same types of crosses but using various members of the *gallica* group the more heterogeneous HYBRID CHINA roses were raised in France and England in the years following 1820.

The second half of the nineteenth century saw the development of two further distinct groups or families of roses, two further steps in mixing. First, there were the diploid TEA roses, dwarf segregates from Noisette and *R. gigantea* crosses. They had flowers of exquisite form, borne all the year round in favourable climates, but were not very hardy. Secondly, there were the tetraploid HYBRID PERPETUALS from Bourbons and Hybrid Chinas crossed back with the European *gallica* varieties. The Hybrid Perpetuals were not as good as their name. They naturally had only a weak second flowering season but they were hardier than the Hybrid Chinas. These two groups included the bulk of the roses grown in the period 1850 to 1890.

From 1890 onwards, the five old groups began to be superseded by a sixth group, the HYBRID TEAS, produced from crosses between these last two and combining the merits of their parents. It was in this group, we may note, that the chromosomes of the three great parent stocks *moschata*, *gallica* and *chinensis* were effectively brought together for the first time, brought together, that is to say, in fertile hybrids. The original Hybrid Teas were again, of course, triploid, but again tetraploids arose by non-reduction in the course of crossing the less sterile triploids back with the Hybrid Perpetuals.

Further variation was brought into the Hybrid Tea group at the beginning of this century from the seventh species, the yellow and copper tetraploid *R. foetida*, in a triple cross. This supposedly Persian species is not known in nature and is itself doubtless, like chinensis, a cultivated group of ancient hybrid origin. Its *bicolor* mutation is responsible for the brilliant yellow, orange, and coppery tones first found in the PERNETIANA group.

Another line of advance began at the same time as the Hybrid Teas. It was purely eastern, purely diploid and therefore less important in its immediate results. It came by crossing the other diploids with the wild *multiflora* of Japan and gave the POLYANTHA roses. These arising from a triple cross in the fourth and later generations were selected in two directions, dwarf and climbing. Unlike the Hybrid Teas, their flowers were very small, and borne in large clusters. The Dwarf Polyanthas, moreover, had a very long flowering season. Many varieties arose later, some of them as bud mutations. The mutation to the formation of pelargonidin came first in the late 1920's to give the diploid Paul Krampel (Scott-Moncrieff, 1936). More recently it gave among its derivatives the tetraploid Sondermeldung or Independence (Dayton, unpub.). Pelargonidin is a pigment unknown in the ancestral species and indeed new for the whole genus.

The most recent group of garden roses, the HYBRID POLYANTHAS, came from crossing the Dwarf Polyanthas with Hybrid Teas. The first varieties were again triploids and were sterile with a few exceptions. From these the later tetraploids have arisen, principally through crossing back to the Hybrid Teas. But there are varieties similar to the true Hybrid Polyanthas in their clusters of medium-sized flowers born over a very long period, and these have *moschata* or *wichuraiana*, in addition to *R. multiflora*, in their ancestry.

Bud Sports. Among these crosses, both $2x$ and $4x$, climbing or rambler varieties have appeared from time to time. They have all been derived from crosses with the four tall or climbing species. Successively the Noisette, Hybrid Tea, and Climbing Polyantha groups have been most important: and last of all Hybrid Wichuraiana have predominated by reason of their disease resistance and longer flowering. One special feature of the climbers is their frequent appearance in the Hybrid Tea and bush roses by bud mutation. Nearly a hundred pairs of bush and climbing forms of Hybrid Tea varieties are named in cultivation. The climbing forms are liable to revert to the bush type often producing intermediate forms. This suggests that they arise by plasmagene rather than nuclear gene mutation.

Another example of bud mutation giving rise to a chimaera is the scarlet rose Paul Krampel which when eight years old gave pink and crimson bud sports on the same plant. Probably the new form had arisen as a chimaera of old and new types and was giving the older pigment types by reassortment of tissues.

The largest scale of bud mutation ever invoked in plants perhaps is that by which double Moss Roses, with resinous hairs on the fruit, according to Hurst, came from the Cabbage

Roses during the eighteenth century. Over sixty clonal varieties are known. All of these are sexually sterile and several of them have been known to revert in cultivation to the Cabbage Rose type. Again, therefore, these mutations may have existed as unstable chimaeras.

The precise internal and external conditions of bud-mutation evidently deserve special study in *Rosa*.

Stocks. In ancient times the two main groups of roses, *gallica* in Europe, *odorata* in China, could be readily propagated by suckers or cuttings. Grafting experiments were no doubt attempted, but it was only when new *gallica* forms arose in the Middle Ages which were difficult to multiply directly, that gardeners would resort to using the borrowed roots of older garden forms without such a defect.

Suitable stocks were no doubt selected in this way by nurseries during the eighteenth century. At this time, too, pentaploid *Caninae* roses began to be used, having the advantage that they could be gathered as wild seedlings. In the nineteenth century three new groups of roses began to be grown and even bred for stocks: (i) the recently introduced diploid species, *multiflora* and *rugosa*; (ii) selections from their hybrid progeny; and (iii) established garden varieties such as *manetti*, a supposed Noisette raised in 1837. The last included not only diploids but also triploids, and tetraploids: differences of chromosome number proved to be no barrier to grafting.

Conclusion. The garden roses have been derived not from the hexaploids and octoploids but entirely from the lower-numbered species. The largest section of the genus, the Cinnamomeae, which contains the greatest development of polyploidy, has taken no part in the great programme of Table 24. The garden forms this section has produced have come by mutation and hybridization from its diploid species, *rugosa* and *cinnamomea*.

Thus almost the whole of rose history has been concerned with raising the progeny of crosses to a uniform tetraploid level. Higher polyploids must certainly have arisen in the course of these innumerable crossings, as they have in *Narcissus*. But they have evidently been rejected by breeders. It is at the tetraploid level that interspecific hybridization has been able to release the great diversity that we now see. It has worked largely by recombination but partly also by the selection of instability. And its future possibilities for the genus are still beyond estimation.

7. *Chrysanthemum*

The importance of diploids as the basis of plant improvement in *Hyacinthus*, *Narcissus* and *Rosa* bears out what we saw of the origins of the staple grain crops. We must now notice two exceptions to this rule, first the Chrysanthemums.

The garden forms have been cultivated in China for over 2,000 years. They are derived, it seems, entirely from hybrids within a hexaploid complex of Chinese species loosely known as *Chrysanthemum indicum* ($6x = 54$). From this complex new wild types of different geographical races have been continually crossed with the main group of ornamentals. New varieties have arisen both as hybrid seedlings and from bud sports: in other words, as recombinants and as mutants. The first effective introduction to Europe was in 1789; the first results of controlled breeding date from 1827, and the first bud sport was recorded in 1832.

The remarkable property of the garden *Chrysanthemum* is that the Japanese and the English varieties, although differing greatly in floral character, show parallel variation in chromosome complement and also in flower size. The bulk have about the hexaploid chromosome number of 54 but they vary from 47 to 64, i.e. from about $5x$ to about $7x$. Furthermore, the larger-flowered varieties, both Japanese and English, have the higher chromosome numbers.

Some of these variations in chromosome number have no doubt arisen in seedlings from irregular meiosis and germ-cell formation in $6x$ parents which have been used for breeding. A seedling with 60 chromosomes may presumably come from parents with 54, or 54 crossed by 60, or indeed from any other known type. But some variations also arise, and are continually arising, from bud-mutations. Chromosomes are lost or gained at mitosis in the meristem in the ordinary course of growth.

Hence layers of tissue must arise with changed chromosome numbers: the variety becomes a chimaera which eventually may grow wholly of the new type; and, apparently, it may be assisted in doing so by the hot water treatment used to destroy pests. The new type survives as a new garden variety because it is commonly of a new and interesting colour and the disturbance in chromosome proportions, the unbalance from a few chromosomes more or less, is tolerated.

In this way 'families' of bud sports grow up, ten or twenty varieties, all derived from a single original seedling. The members of the family differ a little in most respects but have certain features (determined by the 45 untouched chromosomes) in common.

Such a method of evolution as that of the ornamental *Chrysanthemum* is not yet known in any other plant. It depends clearly on not too high a degree of differentiation of the 9 chromosomes in the basic set. Each chromosome has, to a greater extent than usual, its own little balance so that changes of number are not eliminated in a hexaploid *Chrysanthemum* as they are in a hexaploid wheat: but it does not show the same disregard of inter-chromosome balance as the garden hyacinth. In the hyacinth a disproportion between chromosomes as great as 3 : 2 is tolerated; in the *Chrysanthemum* no disproportion is greater than 6 : 5.

8. *Dahlia variabilis*

The garden dahlia was introduced to Europe as a tuber in 1789 and as seed in 1804. But it had been bred and grown for ornament in Mexico even in pre-cortesian times and had produced its full range of colour before it reached Europe.

The wild species from which *D. variabilis* ($8x = 64$) is supposed to have arisen are tetraploids with 32 chromosomes; but they are allo-tetraploids and there are no diploid species. The parents of *variabilis* on biochemical, chromosomal and breeding evidence are believed to have been tetraploid species of two groups whose chromosomes however readily pair. One of these, *imperialis*, has as pigments a magenta anthocyanin and an ivory flavone; the other, *coccinea*, has a scarlet anthocyanin and a yellow chalkone. The result of their union or, we may say, their addition, has been to give in the octoploid the whole of their range of flower colours in all possible combinations.

These combinations owe their extraordinary wealth of diversity not to any innovation but to mere reshuffling. This was made possible because the octoploid retained free pairing among the 4 chromosomes of each of 16 kinds. It is thus an auto-tetraploid in relation to its allo-tetraploid parents. It has been able to do this and still retain its fertility owing to a strict localization of crossing-over and chiasmata near the ends of the chromosomes as happens in ring-forming plants like *Oenothera*. At the same time the octoploid seems to be strictly self-incompatible. This property must have been handed down to it from its wild ancestors notwithstanding that the tetraploid species we now have are self-fertile. The incompatibility has hastened the very process which was necessary to produce recombination at a time when breeders would use natural open-pollinated seed.

VII. LESSONS OF CHROMOSOME BOTANY

CHROMOSOME Botany, as we have now seen it, has proved to be chiefly effective in showing evolutionary principles. The materials have fallen into three groups in regard to the type of evidence and the scale of evolution we have been examining. Wild plants, the staple and ancient economic plants, and the ornamental plants have been undergoing evolutionary changes over three ranges of time, roughly speaking over millions, thousands and hundreds of years. The evidence, therefore, is of different kinds and degrees of cogency. Experimental method has an increasing relevance as we move down the scale, as we pass from the evolution of the most ancient trees to recent improvements in roses and daffodils, as we pass from historic to contemporary evolution.

If the three compartments were indeed water-tight their comparison might be highly speculative. But the chromosome approach has in fact joined the three together. It has done so for the unexpected reason that different plants in each field are seen to evolve at entirely different speeds on the chromosome scale with respect to the external or classical scale. The chromosomes change as though controlled by a three-speed gear-box. Which, indeed, is true, since number, structure and gene content vary independently with different orders of effects. In some plants, therefore, the chromosomes change visibly in direct relation with the external symptoms of evolution. In others they move much faster; and in others again they are hardly seen to move at all.

The relation which we noted of chromosome stability to the length of the life cycle in large systematic groups implies that the life cycle is itself fairly stable. This might seem surprising since the higher plants, unlike the higher animals, can by slight genetic variations contract or expand their life cycles. Thus a tropical perennial like wild cotton, as we saw, can quickly turn itself into a temperate annual. Or an annual can delay its maturity and develop storage organs. As we noticed in countless polyploids this allows it to extend its life cycle without limit, to dispense with sexual fertility, and even totally replace sexual by asexual reproduction. Probably, however, these observations are misleading. The shortening of the life cycle may be rare, apart from the artificial selection of cultivated plants, in face of the quicker adaptability of quicker growing competitors. And the lengthening of the life cycle will, we have reason to expect, reduce adaptability and lead into an evolutionary blind alley.

The connection between stability and life cycle is in fact confirmed by the general study of cultivated plants. And for the ornamental plants it even becomes a documented and experimental fact which can be proved or disproved by the chromosome numbers. The chromosomes indeed act as a cross-referencing system checking our other evidence whether it is on the hundred-year scale or the million-year scale.

When this system of checking and testing is applied to the ornamental plants it demonstrates a number of principles on the strictly genetic as well as the evolutionary level. First, cultivation by itself has never had a genetic effect on a cultivated plant. It has repeatedly failed to create variability or to direct variation. Secondly, selection can work very rapid effects provided it is given variation to work on. The effect of cultivation is that larger growing plants will survive if the genotypes for larger growth are available. Thirdly, new variation arises only from a change in the breeding system, either from the crossing of inbred species or from the prevention of crossing in outbred species. Finally, the parts played in creating variation by gene-mutation, gene-rearrangement, chromosome unbalance and new or old polyploidy are radically different in different groups. This has been shown by many joint systematic, genetical and cytological studies: the type of variation depends on the genetic organization of the chromosomes with which nature or the plant breeder sets to work.

A comparison of chromosome variation in different natural groups now provides unlimited evidence of this principle. The importance of polyploidy in some groups, its absence in others, is clear. Studies of heterochromatin show just the same. The New World species of *Fritillaria* are all well stocked with heterochromatin: the Old World species have hardly any at all. Again, from meiosis in hybrids, as we saw, in some genera all species differ by inversions of segments: but in others we find no evidence of inversions at all. Probably a localization of crossing-over achieves the same effect of holding genes together in blocks.

These differences in the variation found in systematic groups depend on differences of genetic organization whose selective action and physiological and mechanical control we only partly understand. But they obviously express themselves in the wide range of character in both natural and artificial evolution that we have now brought to light. Within this range of variation one contrast is most instructive. On the one hand, there are those groups where variation is immediately released by hybridization or by polyploidy or by both. This is a situation which is easy to understand and universally admitted. On the other hand, there are those groups where variation arises not from crossing but rather from inbreeding, not from polyploidy but in a strict diploid. The variation often arises, not at once, but after some delay. Here several questions are still unanswered. What happens during this delay? How does instability eventually arise? Could we by any means induce it without waiting so long? These are questions that have long interested plant breeders and evolutionists.

The cases of *Hyacinthus*, *Lathyrus* and *Primula sinensis* make the situation much clearer. In these plants a great burst of variation can arise after some generations of breeding from a single plant or a single type. Presumably changes arise from errors in sexual reproduction, especially no doubt irregularities of crossing-over at meiosis, which would never arise from vegetative propagation. Such irregularities are well known in experimental observations on maize and rye when the outbreeding species is compelled to inbreed. And, as we have seen, this is just what happens when representatives of cross-fertilizing species are compelled by isolation, or by introduction, or by migration, to fertilize themselves. The mutants that arise in this way are not only interesting in themselves: they are also interesting because they have been selected for instability or mutability, a principle which has been demonstrated experimentally by Harland. Their instability is therefore cumulative.

This principle was understood, although not in the precise terms in which we can put it, by the most experienced and acute plant breeder of the last century. This is how Darwin puts it (1868, Vol. 2, p. 262):

The most celebrated horticulturist in France, namely Vilmorin, even maintains that, when any particular variation is desired, the first step is to get the plant to vary in any manner whatever, and to go on selecting the most variable individuals, even though they vary in the wrong direction; for the fixed character of the species being once broken, the desired variation will sooner or later appear.

Today, of course, we have a substitute for waiting until a destabilizing mutation appears: we can induce mutations by treatment with X-rays. However it comes about, the result is clear. From a homogeneous group a heterogeneous group may be produced, a group within which one may now speak not only of crossing but of 'hybridization'. Such hybridization leads to recombination from which a variety of forms become available for the selection of the breeder.

This series of events is derived from the observation of ornamental plants over a period of only a few hundred years. Let us recall that Darwin used the more recent improvement of cultivated plants, in general, to throw light on the more remote and less accessible problem of evolution as a whole. In the same way we are using the more recent improvement of ornamental plants to throw light on the more remote problem of the origin of cultivated plants as a whole. In this way we learn what must have happened to certain ancient economic plants of obscure origin such as *Vicia faba* and *Zea mays* over a period of a few thousand years, partly in producing diversity and partly in shifting the whole character of the species in the direction required by man. Already, except in very tenacious woody perennials like roses, the modern varieties have often forced out of cultivation their outmoded predecessors which would otherwise have revealed to us their ancestry and their evolution both in outward form and inner chromosomes.

The explanation of the principle to be deduced from the

pioneer work in chromosome geography discussed earlier is at the same time disclosed by the ornamental plants: the principle that instability is favoured on the periphery or colonizing margin of a species. It is that, on the margin, selection will always favour change. At the same time local isolation will favour the breakdown of the breeding system in such a way as to provide the genetic basis of change. The internal and the external pressures agree in promoting evolution. For this reason evolutionary change is correlated with territorial movement.

From our chromosome studies of wild and cultivated plants we learn that changes in the breeding system are crucial in evolution and that they operate in two stages. Primarily, crossing is favoured by itself favouring recombination. Secondarily, the hybridity that crossing produces becomes an end in itself. A change towards selfing or towards wider crossing disturbs both these adjustments. The reaction of the species to these disturbances is to break up all balanced polymorphisms, that is to break up the system which has held the species together and so to set in motion the variety of evolutionary processes we find in both wild and cultivated plants.

APPENDIX I

EARLIEST USE IN ENGLISH OF NAMES FOR CULTIVATED PLANTS

(Oxford English Dictionary)

O E	Wheat	Beet	Lily	Mulberry
	Barley	Radish	Rose	Walnut
	Oats	Leek		
	Rye	Garlic	Parsley	Apple
	Pease	Asparagus	Mint	Pear
	Clover			Plum
		Hemp		Cherry
		⌠Flax		
	Bean	⌡Linseed		Strawberry

M E	Millet	Parsnip	Lettuce	Quince
		Onion	Endive	Medlar
			Cucumber	Bullace
		Cabbage		Damson
		Kale	Rhubarb	Fig
		Mustard	Laurel (Bay)	Peach
	⌠Vetch	Rape	Hop	Grape
	⌡Tare			Orange

1500		1570		1660	
Chestnut	1519	Tobacco	1577	Celery	1664
Spinach	1530	Tulip	1578	Pineapple	1664
Artichoke	1531			Broccoli	1699
Turnip	1533	Cauliflower	1597	Sweet Pea	1732
Carrot	1533	Potato	1597		
		Banana	1597	Brussels sprout	1748
Buckwheat	1548				
Kidney Bean	1548	Tomato	1604	Mangel wurzel	1779
Apricot	1551	Lilac	1625	Kohlrabi	1807
Horse Radish	1561	Sainfoin	1625		
Maize	1565	Lucerne	1626	Swede (1781)	1812

213

APPENDIX II

EVOLUTIONARY PROCESSES IN ANIMALS

I HAVE had the pleasure of reading this book in proof, and was impressed with the fact that the principles which it develops, or clarifies, for the first time, are clearly applicable to animals as well as to plants. For these concepts illuminate some of the obscurities not only of botany but of biology as a whole, and in a way which will transform certain aspects of evolutionary study and of taxonomy. It was for this reason that I made a few notes upon some further extensions of them, and upon one or two of their zoological bearings, for animals and plants have sometimes had to solve the same problems by different means.

It is essential for certain species to develop at varying rates, so that a proportion of the individuals may escape an unfavourable season. This is achieved in some plants, *Nicandra physaloides* for instance, by retaining or losing one member of a pair of iso-chromosomes. The plants with 20 chromosomes germinate first, while those with 19 may germinate after many years, the populations thus showing a chromosome dimorphism (p. 14). In the Lepidoptera (Moths), a similar insurance is sometimes obtained by varying the length of the pupal period, a form of variation that has been little studied, but is known to be under genetic control. Normally all the pupae of a brood hatch at the same season. However, in those northern insects which fly early in the spring and are, therefore, liable to be killed by inclement weather, the emergence is scattered over a period of several years: for example, in the large Agrotid moth *Brachionycha nubeculosa* Esp., which is to be found in restricted localities in the Scottish highlands in late March and early April. A similar situation is found in coastal moths (for example, *Agrotis vestigialis* and *Euxoa cursoria*) which pupate in sand. Their pupae may be temporarily exposed, or buried to a great depth, due to high winds shifting the surface of the dunes: thus a variable pupal period is something of a safeguard to such species.

Genetic isolation is much more effective than the geographical type which, in animals as in plants, acts capriciously (p. 38). In plants, genetic isolation can arise through chromosome changes, and there is no doubt that it may do so in animals also, for related species have chromosome numbers which clearly indicate polyploidy, with or without some minor numerical adjustments. Thus, as Darlington points out to me, in Orthoptera, Hymenoptera and Lepidoptera what are apparently diploid and tetraploid species are known within a genus. Furthermore, the only cross within the various 'sub-species' of the butterfly *Papilio machaon* L., which has led to a deficiency of the heterogametic sex (the female in the Lepidoptera), is that between the European *P. m. britannicus* and the American *P. m. hippocrates* (Clarke and Sheppard, 1955). Macki and Makino (1953) give the haploid number of the latter species as 31, while Lorkovic (1941) reports that $n = 30$ for European sub-species of *machaon* and, according to Federley (1939), n varies from 30 to 33 in the homogametic sex of these forms.

The principle of mutual selection (p. 47) is one of the utmost importance, which has received exceedingly little precise study. The evidence for it, though indirect, is cogent. It is principally derived from an aspect of polymorphism which chances to have been developed in investigating animal rather than plant material. That is to say, the discovery that polymorphism is with great frequency maintained by a group of two or more closely linked genes, which act as a unit in controlling alternative forms. For if genes interact advantageously, selection will evidently operate to produce, and then to increase, linkage between them (Fisher, 1930; Sheppard, 1953). It is an impressive fact that instances of polymorphisms associated with close linkage, often first reported as multiple allelomorphs or the multiple effects of single genes, are known from a wide variety of organisms, among which may be cited (examples only): Orthoptera (*Apotettix eurycephalus* and *Paratettix texanus*), snails (*Cepaea nemoralis* and *C. hortensis*), fish (*Lebistes reticulatus*), and several of the human blood groups. Also polymorphic butterflies in which it is essential that very different characters, such as colour-pattern and structure, should segregate together in order to produce a given mimetic resemblance in one phase only (e.g. *Papilio memnon*). It has been suggested that an entirely different process is responsible for this occurrence: that is to say, duplication of the gene, with mutation taking place in one of the loci, but it is likely that such favourable interactions will tend to occur most often between different genes controlling the same processes. Moreover, there are instances of polymorphism maintained by such closely linked genes which, however, have very distinct effects, indicating that they have been brought together, rather than duplicated. A good instance of that situation is provided by the mimetic butterfly *Papilio memnon*, in which the units of a super-gene control colour-pattern and the presence or absence of tails on the hind wings. Therefore pigmentation and wing-shape segregate together in a block, giving an appropriate resemblance to highly distinct models one of which is tailed and the others tailless. Rare cross-overs have made it possible to determine the order of the genes concerned (*cf.* heterostyly, p. 48).

It is, of course, inherent in the concept of mutual selection that genes can interact to produce new effects. This property is substantiated by many striking instances, but the subject has been studied very little in comparison with its importance. Two animal examples may usefully be mentioned: (1) The gene responsible for dominant black spotting in the fish *Platypoecilus* gives rise to fatal melanotic tumours in the F_1 hybrids with *Xiphophorus*, though such hybrids, when lacking this gene, are healthy. (2) The hybrids between Pearlneck doves and Ringdoves contain antigens absent from both parental forms.

Darlington points out that species may be exceptional at the edge of their range, for he stresses that 'different requirements of soil, and of climate, and of relationships with other organisms, push the diverging stocks or races, if they are genetically isolated from one another, into different regions. These in turn becoming the centres of new diversity' (p. 62). He further remarks that, 'instability is favoured on the periphery or colonizing margin of a species' (p. 167).

With this I am in complete agreement, but I would extend the concept in another direction. Species at the edge of their range can be colonizing, which is the situation discussed in this book, or they may survive with difficulty in an environment which has reached the limit of their toleration. The latter situation induces a different result for, in order to maintain themselves at all in such unfavourable conditions, it is necessary for organisms to select some type of habitat to which they can become closely adapted. It is for this reason that many widely tolerant species are at the edge of their range wholly confined by ecological requirements to which they are not restricted elsewhere. It is worth illustrating this concept with a few examples.

The moth *Malacosoma castrensis* L. reaches the extreme north west of its range in Britain, where it only occurs in south eastern England. There it is restricted to coastal salt marshes, though it is widespread in Continental Europe, where it inhabits woodlands, heath, and other types of country, and has no special connection with the shore. It feeds upon a wide variety of common plants, so that its habitat is not thus limited in England by the distribution of its larval food. Accordingly, in this country, but not elsewhere, it has achieved an associated physiological adaptation; the eggs have become resistant to sea water. They are dispersed by being laid in batches on floating debris which is washed round the salterns and stranded at high tides.

A comparable situation is found in two butterflies. *Melitaea cinxia* L., which feeds upon *Plantago* species, is entirely restricted to the southern coast of the Isle of Wight, and *Thymelicus acteon* Rott., a grass feeder, to a small coastal strip in Dorset and Devon. Neither is found more than a mile from the sea, but they are both widespread in central Europe.

Mr. M. H. Williamson points out to me that the Harvest Mite *Nelina sylvatica* Sim. (Class Arachnida, Order Opiliones) is almost entirely coastal in Britain, though it has once or twice been found by lakes and rivers in Ireland. It has, however, an extensive distribution in Europe, occurring especially in woods.

A number of snails, *Ena montana* Drap., *Helix pomatia* L., and others, are confined to calcareous soils in England, but are indifferent to the presence of lime on the Continent. Presumably in the difficult conditions imposed upon them at the north western edge of their range, they require a better supply of lime for shell formation than is necessary for them elsewhere.

It is obviously important that species surviving at the edge of their range by close adaptation to special conditions, should be restricted in their dispersal, so that they neither interbreed with the more tolerant forms nearer the centre of their distribution, nor stray from the conditions to which they are adjusted. An example illustrating this proposition is provided by the Swallowtail butterfly, *Papilio machaon* L., which maintains itself in Britain only in eastern England. Here it has adapted itself as a purely marshland species, inhabiting a number of the Norfolk Broads, where it feeds only upon *Peucedanum palustre*. In Continental Europe, it is tolerant of a wide range of habitats, lowland to Alpine, damp to dry, and feeds on numerous *Cruciferae*. It is there a wandering insect, but in Britain its habits are quite different; for it is extraordinarily localized and, though powerful on the wing, never strays far from the marshes where it is adjusted to live.

One further point arises in this connection. Individuals at the colonizing margin of the species should clearly maintain a high variability. On the other hand, once they have evolved adaptations which enable them to live successfully at the extremity of the conditions which they can tolerate, they should vary as little as possible. For example, *Primula scotica* is restricted to the furthest limits of Britain: the far north of Scotland, and some of the northerly Scottish islands. It is probably no coincidence, therefore, that this species is the only homostyled member of the British *Promulaceae*. It has obviously been derived from *P. farinosa*, as an extreme specialization to meet the conditions at the edge of its range, by a series of adaptations which include habit, ecological requirements, polyploidy, and a floral structure which promotes inbreeding.

Indeed, selection is at the present time favouring the spread of homostyles in two populations of the primrose, illustrating the way in which one aspect of such speciation as that which gave rise to *P. scotica*, can take place. Furthermore, such populations, passing as they are from predominantly outbreeding to predominantly inbreeding conditions, provide us with experimental material in which it might be profitable to search for the type of evolution exemplified by *Erophila verna* (p. 73).

E. B. FORD

REFERENCES

1. Clarke, C. A., and Sheppard, P. M., 1955: *Evolution* **9**, 182–201.
2. Federley, H., 1939: *Hereditas* **24**, 397–464.
3. Fisher, R. A., 1930: *The Genetical Theory of Natural Selection*. Oxford.
4. Lorkovic, Z., 1941: *Chromosoma* **2**, 155–191.
5. Macki, K., and Makino, S., 1953: *Lepidopterist's News* **7**, No. 2.
6. Sheppard, P. M., 1953: *Am. Nat.* **87**, 283–294.

NOTE

For a general bibliography of chromosome numbers refer to the companion volume, C. D. Darlington and A. P. Wylie, *Chromosome Atlas of Flowering Plants* (Allen & Unwin).

BIBLIOGRAPHY

CHAPTER I

AYONOADU, U. W. and REES, H., 1968. The Regulation of Mitosis by B-Chromosomes in Rye. *Exp. Cell Res.* **52**, 284–90.

BARLOW, P. W. and VOSA, C. G., 1970. The Effect of Supernumerary Chromosomes on Meiosis in *Puschkinia. Chromosoma* **30**, 344–55.

CATCHESIDE, D. G., 1950. The B-chromosomes of *Parthenium argentatum. Genet. Iber.* **2**, 139–148.

DAKER, M. G., 1967. A Haploid Cultivar of *Pelargonium*, etc. *Chromosoma* **21**, 250–71.

DARLINGTON, C. D., 1941. Polyploidy, Crossing-over, and Heterochromatin in *Paris. Ann. Bot.* n.s. **5**, 203–216.

—— 1965. *Cytology.* (3rd ed. Recent Advances in Cytology.) Churchill, London.

—— and JANAKI AMMAL, E. K., 1945. Adaptive Isochromosomes in *Nicandra. Ann. Bot.* n.s. **9**, 267–281.

—— and LA COUR, L. F., 1941. Nucleic Acid Starvation of Chromosomes in *Trillium. J. Genet.* **40**, 185–213.

—— —— 1950. Hybridity Selection in *Campanula. Heredity*, **4**, 217–248.

—— and SHAW, G. W., 1959. Parallel polymorphism in the heterochromatin of *Trillium* species. *Heredity*, **13**, 89–121.

—— and THOMAS, P. T., 1943. Morbid Mitosis and the Activity of Inert Chromosomes in *Sorghum. Proc. Roy. Soc. B. B.* **130**, 127–150.

—— and UPCOTT, M. B., 1941. The activity of inert chromosomes in *Zea Mays. J. Genet.* **41**, 275–296.

DUNCAN, R. E., 1945. Variable Aneuploid Numbers of Chromosomes in *Paphiopedilum wardii*, etc. *Amer. J. Bot.* **32**, 506–509.

FERNANDES, A., 1952. Sur le Role probable des Hétérochromatinosomes dan l'Evolution des Nombres Chromosomiques. *Sci. Gen*, **4**, 168–181.

HÅKANSSON, A., 1948. Behaviour of Accessory Rye Chromosomes in the Embryo Sac. *Hereditas*, **34**, 35–59.

JACKSON, R. C., 1962. Interspecific hybridization in *Haplopappus*, etc. *Amer. J. Bot.* **49**, 119–132.

LEWIS, H., 1951. The Origin of Supernumerary Chromosomes in Natural Populations of *Clarkia elegans. Evolution*, **5**, 142–157.

LÖVE, A. and LÖVE, D., 1944. Cyto-taxonomical Studies on Boreal Plants, III. *Ark. Bot.* **31**, No. 12, 1–22.

MAUDE, P. F., 1940. Chromosome Numbers in some British Plants. *New Phytol.* **39**, 17–32.

MÜNTZING, A., 1948. Accessory Chromosomes in *Poa alpina. Heredity* **2**, 49–61.

MÜNTZING, A., 1949. Accessory Chromosomes in *Secale* and *Poa. Proc. 8th Int. Congr. Genet.* Stockholm. *Hereditas, Lund.* suppl. **35**, 402–411.

—— 1950. Accessory Chromosomes in Rye Populations from Turkey and Afghanistan. *Abstr. Hereditas, Lund.* **36**, 507–509.

—— 1968. A New Category of Chromosomes. *Proc. 10th Int. Cong. Gen.* **1**, 453–467.

NAVASHIN, M., 1934. Chromosome Alterations caused by Hybridization and their Bearing upon Certain General Genetic Problems. *Cytologia.* Tokyo, **5**, 169–203.

NODA, S., 1968. Achiasmate bivalent formation by parallel pairing in PMC's of *Fritillaria amabilis. Bot. Mag. Tokyo* **81**, 344–5.

POHLHEIM, F., 1968. Thuja gigantea gracilis—ein Haplont unterden Gymnospermen. *Biol. Rundschau* **6**(2).

ROMAN, H., 1948. Directed Fertilization in Maize. *Proc. Nat. Acad. Sci.* **34**, 36–42.

—— 1950. Factors affecting non-disjunction in Maize. *Genetics* **35**, 132.

RUTISHAUSER, A., 1960. Zur Genetik überzähliger Chromosomen. *Arch. J. Klaus. Stift.* **35**, 440–458.

—— 1969. *Embryologie und Fortpflanzungs-biologie der Angiospermen.* Springer, Vienna.

UPCOTT, M., 1939. The Nature of Tetraploidy in *Primula kewensis. J. Genet.* **39**, 79–100.

VOSA, C. G., 1966. Seed germination and B-chromosomes in the leek (*Allium porrum*). *Chromosomes Today* **1**, 24–7.

—— 1971. The Quinacrine Fluorescence Patterns of the Chromosomes of *Allium carinatum. Chromosoma* **33**, 382–5

—— and BARLOW, P. W., 1972. Meoisis and B-chromosomes in *Listera ovata* (Orchidaceae). *Caryologia* **25** (1), 1–8.

WYLIE, A. P., 1952. The History of the Garden Narcissi. *Heredity* **6**, 137–156.

CHAPTER II

BABCOCK, E. B., 1947. *The Genus Crepis.* Berkeley, Calif.

—— and JENKINS, J. A., 1943. Chromosomes and Phylogeny in *Crepis*, III. *Univ. Calif. Publ. Bot.* **18**, 241–291.

BARLOW, B. A., 1958. Heteroploid twins and apomixis in *Casuarina nana. Austral. J. Bot.* **6**, 204–219.

—— 1959. Polyploidy and apomixis in the *Casuarina distyla* species group. *Austral. J. Bot.* **7**, 238–251.

BROCK, R. D., 1953. Species Formation in *Trifolium subterraneum. Nature*, **171**, 939.

CLAUSEN, J., KECK, D. D. and HIESEY, W. M., 1941. Regional Differentiation in Plant Species. *Amer. Nat.* **75**, 231–250.

CRANE, M. B. and THOMAS, P. T., 1949. Reproductive Versatility in *Rubus. Heredity* **3**, 99–107.

CROWE, L. K., 1964. The evolution of outbreeding in plants. I. *Heredity* **19**, 435–57.

DARLINGTON, C. D., 1940. *Taxonomic Species and Genetic Systems. The New Systematics.* Oxford, 137–160.

—— 1958. *The Evolution of Genetic Systems*. Edinburgh.

DARWIN, C., 1877. *The Different Forms of Flowers on Plants of the same Species*. Murray, London.

FINCH, R. A., 1967. Natural chromosome variation in *Leontodon*. *Heredity* **22**, 359–86.

FORD, E. B., 1957. Polymorphism in plants, animals and man. *Nature* **180**, 1315–1319.

GRANT, VERNE, 1958. The regulation of recombination in plants. *Cold Spr. Harb. Symp. quant. Biol.* **23**, 337–363.

GUSTAFSSON, A., 1946–7. *Apomixis in Higher Plants*. Lunds Univ. Arsskr. N.F. 2. 42(3), 43 (2, 12).

—— 1948. Polyploidy, Life Form and Vegetative Reproduction. *Hereditas* **34**, 1–22.

HAGERUP, O., 1940. Studies on the Significance of Polyploidy, IV, *Oxycoccus*. *Hereditas* **26**, 309–410.

HAIR, J. B., 1956. Subsexual reproduction in *Agropyron*. *Heredity* **10** 129–160.

HESLOP-HARRISON, J., 1953. *New Concepts in Flowering Plant Taxonomy*. London.

KAPPERT, H., 1961. Uber einen Fall von Koppelungswechsel und permanenter Strukturheterozygotie bei *Matthiola incana* *Zuchter*, **31**, 187–192.

LAWRENCE, W. E., 1947. Chromosome Numbers in *Achillea* in Relation to Geographic Distribution. *Amer. J. Bot.* **34**, 538–545.

LEWIS, D., 1942. The evolution of sex in flowering plants. *Biol. Rev.* **17**, 46–67.

LÖVE, A., 1957. Sex determination in *Rumex*. *Proc. Gen. Soc. Canada* **2**, 31–36.

MAUDE, P. F., 1939. The Merton Catalogue. A List of the Chromosome Numbers of Species of British Flowering Plants. *New Phytol.* **38**, 1–31.

NAGAO, S. and MASIMA, I., 1943. Somatische Chromosomen von *Chelidonium majus. Trans. Sapporo Nat. Hist. Soc.* **17**, 132–139.

NYGREN, A., 1957. A fertile Hybrid *Lychnis flos cuculi* and *Melandrium* and its sex segregating offspring. *Kungl. Landtbr. Akad.* **23**, 413–421.

REES, H., 1955. Genotypic Control of Chromosome Behaviour in Rye. *Heredity* **9**, 93–116.

—— 1961. Genotypic control of chromosome form and behaviour. *Bot. Rev.* **27**, 288–318.

RUTISHAUSER, A., 1960. Untersuchungen über die Evolution pseudogamer Arten. *Ber. Schw. Bot. Ges.* **70**, 113–125.

SASISAKA, M., 1954. Chromosome numbers in relation to plant habit. *Proc. 7th Int. Bot. Congr.* (1950), 286–287.

SMITH, B. W., 1955. Sex chromosomes and natural polyploidy in dioecious *Rumex. J. Hered.* **46**, 226–232.

SØRENSEN, T. and GUDJONSSON, G., 1946. Spontaneous chromosome aberrants apomictic *Taraxaca. Kong. Dansk Vid. Selsk. Biol. Skr.* **4** (2), cf. Darlington and Mather, 1949, Figs. 67 and 68.

UPCOTT, M. B. and PHILP, J., 1939. The Genetic Structure of *Tulipa*, IV. *J. Genet.* **38**, 91–123.

VED-BRAT, S., 1965. Genetic Systems in *Allium* III. Meiosis and Breeding Systems. *Heredity* **20**, 325–39.

WATANABE, H., 1962. An X-ray-induced strain of Ring-of-12 in *Tradescantia paludosa*. *Nature* **193**, 603.

WESTERGAARD, M., 1958. The Mechanism of Sex Determination in Dioecious Flowering Plants. *Advances Genet.* **9**, 217–281.

WIMBER, D. E., 1968. Bivalent and ring-forming *Rhoeo discolor* and their hybrids. *Am. J. Bot.* **55**, 572–4.

CHAPTER III

BALDWIN, J. T. and SPEESE, B. M., 1951. *Penthorum sedoides*. *Bull. Torrey Bot. Club*, **78**, 70, 161, 254. *Am. J. Bot.* **38**, 153.

BANACH, E., 1950. Studies in Karyological Differentiation of *Cardamine pratensis* L. in connection with ecology. *Bull. Acad. Pol.*, Ser. B. (1), 197–211.

BARBER, H. N., 1941. Evolution in the Genus *Paeonia*. *Nature*, **148**, 227.

CLELAND, R. E., 1949. Phylogenetic Relationships in Oenothera *Proc. 8th Int. Congr. Genet.*, Stockholm. *Hereditas*, supplt. **35**, 173–188.

DARLINGTON, C. D. and MATHER, K., 1949. *The Elements of Genetics.* London.

GOODSPEED, T. H., 1945. Studies in Nicotiana, III. A Taxonomic Organization of the Genus. *Univ. Calif. Publ. Bot.* **18**, 335–344.

—— 1945. Chromosome Number and Morphology in Nicotiana, IV. Karyotypes of Fifty-five Species in Relation to a Taxonomic Revision of the Genus. *Univ. Calif. Publ. Bot.* **18**, 345–368.

GREGOR, J. W., 1944. The Ecotype. *Biol. Rev.* **19**, 20–30.

HAIR, J. B. and BEUZENBERG, E. J., 1961. High polyploidy in a New Zealand Poa ($38x = 264$). *Nature* **189**, 160.

HOWARD, H. W., 1948. Chromosome number of *Cardamine pratensis*. *Nature* **161**, 277.

JAMES, S. H. 1965, 1970. Complex hybridity in *Isotoma petraea*. I. *Heredity* **20**, 341–53; II. *Heredity* **25**, 53–77.

JANAKI AMMAL, E. K., 1950. Polyploidy in the Genus Rhododendron. *The Rhododendron Year Book*, London, 92–98.

—— 1954. The Cyto-geography of the Genus Buddleia in Asia. *Science and Cult.*, **19**, 578–581.

KIELLANDER, C. L., 1942. A Subhaploid *Poa pratensis* L. with 18 Chromosomes and its Progeny. *Svensk. Bot. Tidskr.* **36**, 200–220.

LEWIS, H., 1953. Quantitative variation in wild genotypes of *Clarkia*. *I.U.B.S. Symp. Gen. Pop. Struc.*, *Pavia*, 114–125.

—— 1953. Chromosome phylogeny and habitat preference of *Clarkia*. *Evolution* **7**, 102–109.

LÖVE, A. and LÖVE, D., 1947. Studies on the Origin of the Icelandic Flora. I. Cyto-ecological Investigations on *Cakile*. *Univ. Inst. Appl. Sci. Reykjavik Dept. Agric. Repts.* Ser. B. 2.

MANTON, I., 1934. The Problem of *Biscutella laevigata* L. Z. *indukt. Abstamm. u. Vererb Lehre*, **67,** 41–57.
—— 1937. The Problem of *Biscutella laevigata* L. II. The Evidence from Meiosis. *Ann. Bot.* n.s. **1,** 439–462.
MOORING, J. S., 1960. A cytogenetic study of *Clarkia unguiculata*. II. Supernumerary Chromosomes. *Amer. J. Bot.* **47,** 847–854.
—— 1961. The evolutionary role of translocations in *Clarkia unguiculata*. *Rec. Adv. Bot.*, **11,** 853–858.
PETO, F. H., 1933. The Cytology of Certain Inter-generic Hybrids between *Festuca* and *Lolium*. *J. Genet.* **28,** 113–156.
SAX, K., 1936. Polyploidy and Geographic Distribution in *Spiraea*. *J. Arnold Arbor.* **17,** 352–356.
SKALINSKA, M., 1950. Studies in Cyto-ecology, Geographic Distribution and Evolution of *Valeriana* L. *Bull. Acad. Pol.* Ser. B. (1) 149–175.
—— 1951. Cyto-ecological Studies in *Poa alpina* L. var. *vivipara* L. *Bull. Acad. Pol.* Ser. B. (1) 253–283.
STAUDT, G., 1951. Uber Fragen der phylogenetischen Entwicklung einiger Arten der Gattung *Fragaria*. *Der Züchter* **21,** 222–232.
STEBBINS, G. L. and BABCOCK, E. B., 1939. The Effects of Polyploidy and Apomixis on the Evolution of Species in *Crepis*. *J. Hered.* **30,** 519–530.
STERN, F. C., 1944. Geographical Distribution of the Genus *Paeonia*. *Proc. Linn. Soc.* **155,** 76–80.
—— 1946. *A Study of the Genus Paeonia*. London.
—— 1949. Chromosome Numbers and Taxonomy. *Proc. Linn. Soc.* **161,** 119–128.
—— 1951. Cytology as a Factor in Classification. *Linn. Soc. Taxon. Pamph.* 57–64.
TURESSON, G., 1930. The Selective Effect of Climate on the Plant Species. *Hereditas* **14,** 99–152.
WALTERS, J. L., 1942. Distribution of Structural Hybrids in *Paeonia californica*. *Am. J. Bot.* **29,** 270–275.
WILMOTT, A. J., 1949. *British Flowering Plants and Modern Systematic Methods*. London.
WINGE, Ø., 1940. Taxonomic and Evolutionary Studies in *Erophila* based on Cytogenetic Investigations. *C.R. Lab. Carlsberg* **23,** 41–74.

CHAPTER IV

ANDERSON, E. and SAX, K., 1936. A Cytological Monograph of the American species of *Tradescantia*. *Bot. Gaz.* **97,** 433–476.
BABCOCK, E. B. and JENKINS, J. A., 1943. Chromosomes and Phylogeny in *Crepis*, III. *Univ. Calif. Publ. Bot.* **18** (12).
BECK, C. (LA COUR, L. F. in), 1953. *Fritillaries*. London.
BENNETT, D., 1972. Nuclear DNA content and minimum generation time. *Heredity* **27,** 390.
DARLINGTON, C. D., 1952. Evolution et Hérédité chez les Plantes. *Ann. Biol.* Paris **28,** 293–300.
—— and LA COUR, L. F., 1940. The Causal Sequence of Meiosis, III.

The Effect of Hybridity in Male and Female Cells in *Lilium*. *J. Genet.* **41,** 49–64.

DARLINGTON, C. D. and MOFFETT, A. A., 1930. Primary and secondary chromosome balance in the *Pyrus*. *J. Genet.* **22,** 130–151.

FERNANDES, A., 1951. Sur la Phylogenie des Espèces du Genre *Narcissus* L. *Bol. Soc. broteria*. **25,** 113–190.

HUTCHINSON, J., 1934. *Families of the Flowering Plants: Monocotyledons*. London.

JANAKI AMMAL, E. K., 1953. The Race History of Magnolias. *Ind. Jour. Genet. Pl. Breed.* **12,** 44–53.

LA COUR, L. F., 1953. The *Luzula* System Analysed by X-rays. *Heredity* **6,** suppl. 77–81.

MCKELVEY, S. D. and SAX, K., 1933. Taxonomic and Cytological Relationships of *Yucca* and *Agave*. *J. Arnold Arbor.* **14,** 76–81.

MATSUURA, H. and SUTO, T., 1935. Contributions to the Idiogram Study in Phanerogamous Plants, I. *J. Fac. Sci. Hokkaido Univ.* Ser. **5,** 33–75.

REES, H. and HAZARIKA, M. H., 1969. Chromosome evolution in *Lathyrus*. *Chromosome Today* **2,** 158–65.

STEBBINS, G. L., *et al.*, 1953. Chromosomes and Phylogeny in the Compositae, Cichoriae. *Univ. Calif. Publ. Bot.* **26,** 401–430.

TANTRAVAHI, R. V., 1971. Multiple character analysis and chromosome studies in the *Tripsacum lanceolatum* complex. *Evolution* **25,** 38–50.

INTERLUDE

DARLINGTON, C. D., 1956. Natural populations and the breakdown of classical genetics. *Proc. Roy. Soc. B.,* **145,** 350–364.

—— 1971a. The Evolution of Polymorphic Systems. *Ecological Genetics and Evolution* (ed. R. Creed). Blackwell, Oxford.

—— 1971b. Axiom and Process in Genetics, *Nature*, **232,** 521–525.

FORD, E. B., 1971. Ecological Genetics (3rd Ed.). Chapman and Hall, London.

GRANT, VERNE, 1971. *Plant Speciation*. Columbia U. P., New York.

CHAPTERS V AND VI

BAKER, H. G., 1965. The evolution of the cultivated Kapok tree. *Inst. Int. Studies, Berkeley*.

BARRAU, J., 1957. L'enigme de la potate douce en océanie. *Et. d'Outre-Mer, Marseille* (April).

—— 1960. Plant Introduction in the tropical Pacific. *Pacific Viewpoint* **1,** 1–10.

—— 1961. Subsistence agriculture in Polynesia and Micronesia. *Bernice P. Bishop Mus. Bull.* 223.

BATESON, W., 1909. *Mendel's Principles of Heredity*. Cambridge.

BATYRENKO, V. G., 1926. Collective data obtained in testing varieties. *Bull. App. Bot. Leningrad* **16** (4), 125.

BAUR, E., 1932. Die Abstammung der Gartenrassen vom Löwenmaülchen. *Züchter* **4,** 57–61.

BEADLE, G. W., 1932. Studies of *Euchlaena* and its Hybrids with *Zea*. *Zeits. ind. Abst. vererb.* **62,** 291–304.

BEALE, G. H., PRICE, J. R. and STURGESS, V. C., 1941. A Survey of Anthocyanins, VII. The Natural Selection of Flower Colour. *Proc. Roy. Soc.* B **130,** 113–126.

BERTSCH, K. and BERTSCH, F., 1947. *Geschichte unserer Kultur-Pflanzen.* Stuttgart.

BRAIDWOOD, R. J., 1960. The agricultural revolution. *Sci. Amer.* **203,** 131–148.

BRUCHER, HEINZ, 1971. Zur Widerlegung von Vavilov's geographisch-botanischer Differential methode (*Ananas, Arachis, Phaseolus*). *Erdkunde.* **25,** 20–36.

BUNYARD, E. A., 1936. *Old Garden Roses.* London.

BURKILL, I. H., 1951. The Rise and Decline of the Yam in the Service of Man. *Adv. of Sci.* **7,** 443–448.

—— 1953. Habits of Man and the Origins of the Cultivated Plants of the Old World. *Proc. Linn. Soc.* session 1964, 12–42.

COLLINS, J. L., 1951. Antiquity of the pineapple in America. *Southwestern J. Anthrop.* **7,** 145–155.

—— 1951. Notes on the origin, history and genetic nature of the Cayenne pineapple. *Pacif. Sci.* **1,** 3–17.

CHANG, K., 1970. Beginnings of Agriculture in the Far East. *Antiquity* **44,** 175–185.

CRANE, M. B., 1940. The Origin and Behaviour of Cultivated Plants. *The New Systematics,* Oxford, 529–547.

—— 1943. The Origin and Relationship of *Brassica* Crops. *J.R. Hort. Soc.* **68,** 172–174.

—— and LAWRENCE, W. J. C., 1952. *Genetics of Garden Plants,* 4th edn. London.

COX, E. H. M., 1945. *Plant Hunting in China.* London.

DARLINGTON, C. D., 1937. *Recent Advances in Cytology,* 3rd edn. London.

—— 1969. *The Evolution of Man and Society,* Allen and Unwin, London.

—— and JANAKI AMMAL, E. K., 1945. Adaptive Isochromosomes in *Nicandra. Ann. Bot.* **9,** 267–281l

—— HAIR, J. B. and HURCOMBE, R., 1951. The History of the Garden Hyacinths. *Heredity* **5,** 233–252.

DARWIN, C., 1868. *Animals and Plants under Domestication.* London.

DOWRICK, G. J., 1952, 1953. The Chromosomes of *Chrysanthemum.* I. The Species. 2. Garden Varieties. *Heredity* **6,** 365–375, **7,** 59–72.

DRUMMOND, J. C. and WILBRAHAM, A., 1939. *The Englishman's Food.* Cape, London.

ENGELBRECHT, TH., 1916. Uber die Entstehung einiger feldmässig angebauter Kulturpflanzen. *Geogr. Z.* **22,** 328–334.

—— 1917. *Über die Entstehung des Kulturroggens.* Festschr. Ed. Hahn. Stuttgart.

FERNANDES, A., 1951. Sur la Phylogénie des Espèces du genre *Narcissus* L. *Bol. Soc. Brot.* **25,** 113–190.

FUSSEL, G. E., 1955. The history of cole (*Brassica* sp.). *Nature,* 9 July.

GALINAT, W. C., 1971. Evolution of Maize and its Relatives. *Ann. Rev. Genet.* **5**.

—— 1971. The Origin of Maize. *Ann. Rev. Gen.* **5**, 447–478.

—— 1972. Maize and Its Relatives. *Adv. Genet.* (In Press).

GRANT, V. and K. A., 1971. Dynamics of clonal microspecies in cholla cactus. *Evolution* **25**, 144–55.

GRIFFITHS, F. P., 1949. Production and Utilization of Alfalfa. *Econ. Bot.* **3**, 170–183.

GUSTAFSSON, A. and SCHRÖDERHEIM, J., 1944. Ascorbic Acid and Hip Fertility in *Rosa* Species. *Nature, Lond.* **153**, 196–197

HARLAND, S. C., 1937. The genetics of cotton. XVII. *J. Genet.* **34**, 158–168.

HAWKES, J. G., 1944. Potato Collecting Expedition in Mexico and South America, II. *Imp. Bur. Pl. Br. and Genet.* Cambridge.

HEHN, X., 1902. *Kulturpflanzen und Hausthiere*, 7th edn. Berlin (trans. by Stallybrass as *The Wanderings of Plants and Animals*, London, 1885).

HEISER, C. B., 1969. *Nightshades* (cultivated Solanaceae). Freeman, San Francisco.

HELBAEK, H., 1959. Notes on the evolution and history of Linum. *KUML: Arb. Jysk. Ark. Selsk.* (1959), 103–129.

—— 1959. Domestication of food plants in the Old World. *Science* **130**, 365–372.

HENDRY, G. W., 1923. The History of Alfalfa. *J. Amer. Soc. Agron.* **15**, 171–176.

HERBERT, W., 1837. *Amaryllidaceae*. London.

HILL, A. W., 1915. The History and Functions of Botanic Gardens. *Ann. Mo. bot. Gdn.* **2**, 185–238.

HORNELL, J., 1946. *Water Transport*, Cambridge U. P.

HURST, C. C., 1941. Notes on the Origin and Evolution of our Garden Roses. *J. R. Hort. Soc.* **66**. I. Ancient Garden Roses (2000 B.C. to A.D. 1800). II. Modern Garden Roses (1800–1940), 73–82; 242–250; 282–289.

HUTCHINSON, J. B., 1954. New evidence on the origin of the old world cotton. *Heredity* **8**, 225–241.

—— 1962. The history and relationships of the world's cottons. *Endeavour* **21**, No. 81, 5–15.

—— 1970. The Genetics of Evolutionary Change (Cotton). *Ind. J. Genet. Plant Br.* **30**, 269–279.

—— SILOW, R. A. and STEPHENS, S. G., 1947. *The Evolution of Gossypium and the Differentiation of Cultivated Cottons*. London.

—— —— —— 1947. *The Evolution of Gossypium*. Oxford U. P., 1947.

JENKINS, J. A., 1948. The origin of the cultivated tomato. *Econ. Bot.* **2**, 379–392.

KENYON, K. M., 1957. *Digging up Jericho*. Benn, London.

—— 1960. Jericho and the Origins of Agriculture. *Adv. Sci.* **17**, 118–120.

KIHARA, H., 1959. The origin of cultivated rice. *Kihara Inst. Biol. Res.* **10**, 68–83.

KIMBER, G., 1961. Basis of the diploid-like meiotic behaviour of polyploid cotton. *Nature* **191**, 98–100.

KRELAGE, E. H., 1946. *Drie Eeuwen Bloembollenexport.* Hague.

KUPTSOV, A. I., 1955. Geographical distribution of cultivated flora and its historical development (Russian). *Bull. All. Union Geog. Soc.* **87**, 220–231.

LAWRENCE, W. J. C., 1931a. The Genetics and Cytology of *Dahlia variabilis. J. Genet.* **24**, 257–306.

—— 1931b. Mutation or Segregation in the Octoploid *Dahlia variabilis. J. Genet.* **24**, 307–324.

LEWIS, D., 1941. The Relationship between Polyploidy and Fruiting Habit in the Cultivated Raspberry. *Proc. 7th Int. Congr. Genet.* Edinburgh. *J. Genet.* suppl. **42**, 190.

—— 1943. The Incompatibility Sieve for Producing Polyploids. *J. Genet.* **45**, 261–264.

LONGLEY, A. E., 1938. Chromosomes of Maize from North American Indians. *J. Agric. Res.* **56**, 107–196.

MANGELSDORF, P. C., 1958. The mutagenic effect of hybridizing maize and teosinte. *Cold Spr. Harb. Sym. quant. Biol.* **23**, 409–421.

—— and REEVES, R. G., 1939. The Origin of Indian Corn and its Relatives. *Texas Agric. Exp. Stn. Bull.* 574.

—— —— 1959. The Origin of Corn. *Bot. Mus. Leaf. Harvard*, **18** (7).

MATHER, K., 1943. Polygenic Inheritance and Natural Selection. *Biol. Rev.* **18**, 32–64.

—— and DE WINTON, D., 1941. Adaptation and Counter-adaptation of the Breeding System in *Primula. Ann. Bot.* n.s. **5**, 297–311.

MENDEL, G., 1865. Reprinted in Bateson, 1909, and in Sinnott, Dunn and Dobzhansky, 1950.

MERRILL, E. D., 1954. The botany of Cook's voyages. *Chron. Bot.* **14**, 161–384.

NISHIYAMA, I., 1959. Studies of the sweet potato and its related species. *Jap. J. Breeding* **11**, 37–43.

—— 1971. Evolution and domestication of the Sweet Potato. *Bot. Mag. Tokyo* **84**, 377–387.

OLSSON, G. *et al.*, 1955. Synthesis of the ssp. Rapifera of *Brassica napus* (swede turnip). *Hereditas* **41**, 241–249.

RASMUSSON, J., 1932. Några undersökningar över *Beta maritima* L. *Bot. Not.* 1932, 33–62.

RICK, C. M., 1958. The role of natural hybridization in the derivation of cultivated tomatoes of Western South America. *Econ. Bot.* **12**, 346–367.

—— and BOWMAN, R. I., 1961. Galapagos tomatoes and tortoises. *Evolution* **15**, 407–417.

—— and NOTANI, N. K., 1961. The tolerance of extra chromosomes by primitive tomatoes. *Genetics* **46**, 1231–1235.

RIDLEY, H. N., 1930. *The Dispersal of Plants throughout the World.* Reeves, Ashford.

RILEY, R. *et al.*, 1958. Evidence on the Origin of the B Genome in Wheat. *J. Heredity* **49**, 91.

RILEY, R., 1960. The Diploidization of polyploid Wheat. *Heredity* **15**, 407–429.

227

RILEY, R. 1966. The genetic regulation of meiotic behaviour in wheat and its relatives. *Hereditas, Sup.* **2**, 396–408.

ROBERTS, H. F., 1929. *Plant Hybridization before Mendel.* Princeton, U.S.A.

ROSANOVA, M. A., 1939. Evolution of Cultivated Raspberry. *C. R. Acad. Sci. U.R.S.S.* **24**, 179–181.

SALAMAN, R. N., 1949. *The History and Social Influence of the Potato.* Cambridge U. P.

SCHIEMANN, E., 1943. Entstehung der Kulturpflanzen. *Ergeb. Biol.* **19**, 409–552.

—— 1951. New Results on the History of Cultivated Cereals. *Heredity* **5**, 305–320.

SCOTT-MONCRIEFF, R., 1936. A Biochemical Survey of some Mendelian Factors for Flower Colour. *J. Genet.* **32**, 117–170.

SEARS, E. R. and OKAMOTO, M., 1957. Intergenomic relationships in hexaploid wheat. *Proc. 10th Int. Cong. Gen.* **2**, 258–259.

SIMMONDS, N. W., 1962. *The Evolution of the Bananas.* Longmans, London.

SINNOTT, E. W. *et al.*, 1950. *Principles of Genetics.* New York.

TÄCKHOLM, G., 1922. Zytologische Studien über *Rosa. Acta Hort. Berg.* **3**, 97–381.

TING, Y. C., 1958. Inversions and other characteristics of Teosinte chromosomes. *Int. J. Cyt.* **23**, 239–250.

TURRAL, J. F., 1965. The old sweet peas. *J. Roy. Hort. Soc.* **90**, 23–29.

VAVILOV, N. I., 1926. Studies on the Origin of Cultivated Plants. *Bull. Appl. Bot.* **16**, 1. **16** (2), 139–248.

—— 1928. Geographische Genzentren unserer Kulturpflanzen. *Verh. 5 Int. Kong. Vererbungswiss.* Berlin, 342–369.

—— 1931a. The role of Central Asia in the origin of cultivated plants. *Bull. Applied Bot. Gen. and Plant Breeding* **26**, 3–44.

—— 1931b. Wild Progenitors of the Fruit Trees of Turkestan and the Caucasus, etc. *Proc. 9th int. Hort. Congr.* London, 271–286.

VAVILOV, N. I., 1935. Theoretical Bases of Plant Breeding. Moscow (Russian).

—— 1949–50. The Origin, Variation, Immunity and Breeding of Cultivated Plants. *Chron. Bot.* **13**, 1–366.

VILMORIN, J. L. DE, 1923. *L'hérédité chez la betterave cultivée.* Paris.

VILMORIN, M. DE, 1951. Cent ans de selection généalogique et cinquante ans de génétique appliquée à Verrières. *Zts. f. Pflanzenzüchtung* **29**, 288–301.

VOSA, C. G., 1969. The hybrid origin of the Tree-Onion. *Chromosomes Today* **2**, 270.

WAINES, J. G., 1971. What do we not know about wheat evolution? *Acta Agron. Ac. Sci. Hung.* **20**, 204–7.

WEATHERWAX, P., 1954. Indian corn in old America. *Quart. Rev. Biol.* **30**, 65–66.

WHITAKER, T. W., 1956. The origin of the cultivated Cucurbita. *Amer. Nat.* **90**, 171–176.

—— and MCCOLLUM, D., 1954. Shattering in lettuce—its inheritance and biological significance. *Bull. Torrey Bot. Club* **81**, 104–110.

WILKES, H. G., 1967. Teosinte: the closest relative of maize. *Bussey Instn. Harvard Univ.*

WILLEY, G. R., 1960. Native New World Cultures, etc. *Evolution of Man* (ed. S. Tax), Chicago.

DE WINTON, D., 1929. Inheritance in *Primula sinensis. J. R. Hort. Soc.* **54,** 84–90.

WYLIE, A. P., 1954a. Chromosomes of Garden Roses. *Amer. Rose Annual,* 1954, 36–66.

—— 1955. The History of Garden Roses. *J. R. Hort. Soc.* **79,** 555; **80,** 87.

YAMASHITA, K. and TANAKA, M., 1960. Einkorn from Syria, Turkey and Greece. *Wheat Inf. Serv.* **11,** 29–30.

ZAGORODSKIKH, P., 1939. New Data on the Origin and Taxonomy of Cultivated Carrot. *C. R. Acad. Sci. U.R.S.S.* **25,** 520–523.

ZOHARY, D., 1959. Is *Hordeum agriocrithon* the ancestor of six-rowed cultivated barley? *Evolution* **13,** 279–280.

—— 1960. Studies on the origin of cultivated barley. *Bull. Res. Counc. Israel* 9D (1), 21–42.

—— and BRICK, Z., 1961. *Triticum dicoccoides* in Israel: Notes on its distribution, ecology and natural hybridization. *Wheat Inf. Serv.* **13,** 6–8.

—— and IMBER, D., 1963. Genetic dimorphism in fruit types in Aegilops speltoides. *Heredity* **18,** 223–231.

ZOSSIMOVICH, V. P., 1939. Evolution of Cultivated Beet, *B. vulgaris. C. R. Acad. Sci. U.R.S.S.* **24,** 73–76.

INDEX

232

234

235